Advanced Innovations in Communication and Computation

Advanced Innovations in Communication and Computation

Edited by

Dr. Ram Krishan

CWP

Central West Publishing

This edition has been published by Central West Publishing, Australia

For more information about the books published by Central West Publishing, please visit https://centralwestpublishing.com

NATIONAL LIBRARY OF AUSTRALIA

A catalogue record for this book is available from the National Library of Australia

ISBN (print): 978-1-922617-44-6

Preface

I am pleased to present you the edited book titled "**Advanced Innovations in Communication and Computation**". In the current era, communication and computing have become intricately connected to provide services to the common man that allow people from all over the world to communicate with ease. As the title suggests, the goal of this book is to evaluate the potential applications of advanced technologies in the field of communications and computation. This book comprises eight chapters contributed by academicians, researchers and students. Chapters in this book mainly cover protocols of wireless communication, antenna design, online learning technologies, mathematical modelling and financial literacy. My family and friends have been very supportive of the publication of this edited book. Finally, I am thankful to all who have contributed and spared their valuable time for this book.

Dr. Ram Krishan
Assistant Professor and Head,
Department of Computer Science,
Mata Sundri University Girls College,
Mansa, Punjab, INDIA

About the Editor

Dr. Ram Krishan presently working as Assistant Professor and Head, Department of Computer Science, Mata Sundri University Girls College, Mansa, Punjab, INDIA (A Constituent College of Punjabi University, Patiala). Dr. Ram obtained his PhD in Computer Science and Engineering from Guru Kashi University, Talwandi Sabo, India in 2017 and M.Tech. in Computer Engineering from Punjabi University, Patiala, India in 2009. He has authored two academic books and has published more than 40 research papers in various international/national journals and conference proceedings, along with book chapters. Dr. Ram has also edited four research books in the field of wireless communication. His research areas include wireless communication, cloud computing and antenna design.

Table of Contents

Chapter 1

Performance Analysis of Fractal Antenna Array Configurations

Sunita Rani[1] and Jagtar Singh Sivia[2]
[1,2]Yadavindra Department of Engineering, Guru Kashi Campus, Talwandi Sabo, Bathinda

Abstract: This paper represents the structure of various fractal array configurations with circular shaped fractals and performance parameters are analysed. For the configuration of fractal array, initially, a circular shaped fractal antenna with partial ground plane is designed, which shows that as we move from 0th iteration to 2nd iteration, the range of wide band is increased. Further linear antenna arrays are designed with 2nd iteration of fractal antenna. Antenna elements are arranged at a distance of $\lambda/2$ with a corporate feed network. Various antenna array structures such as 1×2 and 1×4 is designed and performance parameters are computed with HFSS software and it has been revealed that as the count of antenna elements is increasing, number of bands are increasing, hence multiband response has been achieved with a peak gain of 10.67 dB, 12.59 dB along with S parameters values -22.0075 dB and -28.2337 dB, respectively. A comparative investigation of concerned parameters shows that antenna array with 4 elements has achieved peak gain of 12.59 dB and it covers operating bands for wireless applications. Fractal array with four patches is fabricated and it has been concluded that measured parameters of proposed antenna array are validated with simulated results.

1. Introduction

Communication system requires antenna as a significant element which is employed for the transmission and reception of electromagnetic waves. Important features of microstrip antennas, such as low profile, low cost, easy to fabricate, has attracted various researchers to design antennas for communication applications. Several procedures are present in literature to provide electromagnetic energy to microstrip patch antennas. These techniques are co-axial probe, aperture coupled and microstrip line feed [1]. Linear and circular polarization can be achieved with microstrip antennas [2]. But there are some limitations which leads to narrow impedance bandwidth and broader radiation pattern with large beam angles and polarization impurity. Consequently, these antennas are not good for point-to-point communication. To recover this limitation of patch antennas, various techniques have been studied and analysed Impedance bandwidth can be enhanced by etching slots of different shapes in radiating patch [3-7]. As telecommunication systems require small and lightweight antennas so, many fractal structures have considered and analysed to decrease the extent of these systems. Fractals create complex structures which are produced through method of iterations [8-9]. It applies a simple procedure which is repetitive numerous times. In the foremost period, Mandelbrot [10] stated a fractal configuration which is widely preferred in designing of antennas. Several forms of fractal fragments had well-defined within literature [11-15]. In spite of these fractal configurations, many fractal designs are created with a square curve [16]. These fractal configurations are based on a self-similar and space-filling criterions [17-18] that use the similar iteration factor at a diverse scale. The choice of ground plane also affects the performance parameters of antenna configurations. The change in dimensions of ground plane provides the variation in radiation features of antenna [19-20]. The effect of ground plane has been studied on performance parameters as directive gain along with half power beam width for a square shape antenna in literature.

A patch antenna is having a narrowband and broad radiation pattern with large beam angles. The antenna array is a collection of identical or non-identical patch antennas, geometrically organized in such a manner so that currents running over them are associated with diverse phases and amplitudes to enhance the radiation characteristics in required direction instead of non-desired direction through the interference of electro-magnetic waves. All the antennas of array are usually similar in shape [21] for practically applications because it becomes easy to design and fabricate. These patch antennas can be of any form (aperture, loops, dipole) [22]. The unique parameter in creating the array is the mutual coupling between the elements of array which is only possible if patch elements are distributed by same distance. The coupling phenomenon also influence the efficiency of the antenna array [23]. Antenna arrays also work for 60 GHz frequency band [24]. For long distance communication highly, directive antennas are required.

Presented work describes the design of a fractal antenna that consists of a circular shape patch with FR-4 substrate material on a partial ground plane. Then fractal concept is applied up to the second iteration. Microstrip line feed is coupled with antenna to provide excitation. Further, two and four elements are arranged to design 1×2 and 1×4 antenna arrays with circular fractals and performance parameters are compared.

2. Proposed Design Configuration

A fractal antenna with circular shape is proposed and simulated by high frequency structure simulator to describe the performance features of proposed antenna. Substrate material FR-4 ($€_r = 4.4$) is employed with a substrate height of 1.6 mm for desired antenna. A microstrip line feed is coupled for impedance matching of designed array. Fractal concept is applied with an iteration factor of 1/3 of original diameter of basic structure. More number of iterations lead to shift in frequency towards lower side. Antenna dimensions of projected structure are computed with subsequent equations (1-2).

$$f_r = \frac{1.8412 v_0}{2\pi a\sqrt{\epsilon_r}} \quad (1)$$

$$a_e = \left\{1 + \frac{2h}{\pi a \epsilon_r}\left[ln\left(\frac{\pi a}{2h}\right) + 1.7726\right]\right\}^{\frac{1}{2}} \quad (2)$$

In stated equations resonant frequency (f_r), speed of light in air (v_0), radius (a) and permittivity (ϵ_r) are defined respectively. Fractal antenna dimensions are stated in Table 1.

Table 1. Antenna design parameters

Parameters	Numerical Values (mm)
Substrate Length	52
Substrate Width	46
Substrate Height	1.6
Radius of Circular patch	15

50 Ω line width	3.0678
Circular patch Diameter (0th Iteration)	30
Circle diameter (1st Iteration)	10, 3.33
Circle diameter (2nd Iteration)	3.33, 1.11

Fractal antenna design starts with a circular patch of diameter 30 mm and a circle of diameter 10 mm is etched in the centre of circle and designed patch is attached with a microstrip feed line of 50 Ω impedance. Design of iteration 0th is shown in Figure 1(a). For the design of first iteration, centre circle of 10 mm is surrounded by four small circles of diameter 3.33 mm which has been displayed in Figure 1(b).

In the last iteration, same process is repeated on the circular patch and each circle of diameter 3.33 mm is surrounded by four circles of diameter 1.11 mm throughout the whole patch. Hence the design for 2nd iteration has been shown in Figure 1(c). The back view of proposed fractal antenna configuration with partial ground plane is illustrated in Figure 1 (d).

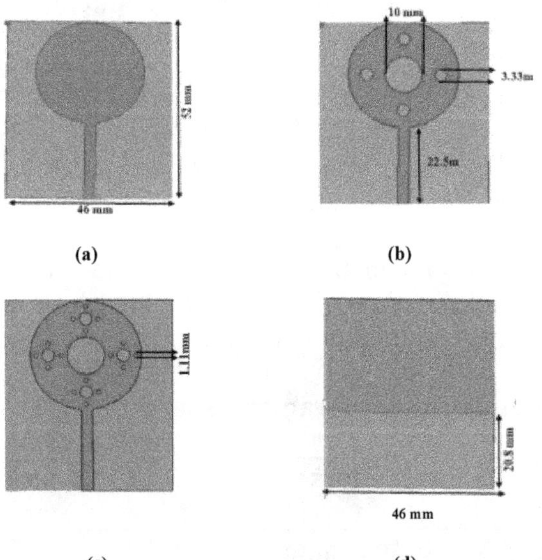

Figure 1. Design of Fractal Antenna (a) 0th iteration (b) 1st iteration (c) 2nd iteration (d) ground plane.

2.1 Antenna Array Design with Two Elements

The circular-shaped fractal array is proposed with FR-4 Epoxy substrate material. Self-similarity property has been employed from 0th to 2nd iteration to design a fractal antenna. Now two circular fractals are positioned at a space of $\lambda/2$ (55.5 mm) away from each other. Antenna array is considered with the combination of two circular fractals along with a unique feeding system. A corporate network of feed lines is used to project a 1×2 linear circular fractal structure of antenna array.

In corporate feeding network a quarter wave transformer is used for matching of 50 Ω feed line impedance and 100 Ω impedance of transmission line. The impedance of the quarter-wave transformer has been calculated in equation 3. Hence width of the quarter-wave transformer depends on its impedance.
Impedance of Microstrip line feed =50 Ω
Impedance of Transmission line = 100 Ω
Quarter wave Transformer impedance = $\sqrt{50 \times 100}$ = 70.7 Ω...... (3)
Dimensions for designing of antenna array with fractals are defined in Table 2.

Table 2. Fractal Antenna Array Specifications

Parameters	Value
Dielectric constant(ϵ_r)	4.4
Height of substrate(mm)	1.6
Radius of Circular patch(mm)	15
50 Ω line width(mm)	3.0678
Quarter wave transformer length(mm)	15.6
Quarter wave transformer width(mm)	1.6
Spacing between two patches	55.5

The proposed antenna array with two fractal patch elements is designed with all these specifications. The design configuration with two elements linear fractal array is presented in Figure 2.

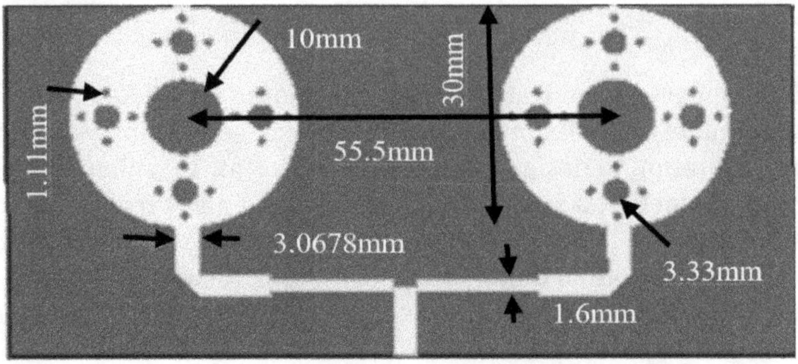

Figure 2. Design of 1×2 Linear Fractal Antenna Array

Designed antenna array is simulated with HFSS software and it is observed that antenna resonates at multiple frequencies hence the response with multiband has achieved.

2.2 Antenna Array Design with Four Elements

The proposed antenna array with four circular fractals is projected with design specification mentioned in Table 2 of section 2.1. In this configuration four patch elements are arranged in a linear manner with an inter element spacing of 55.5 mm as illustrated in Figure 3.

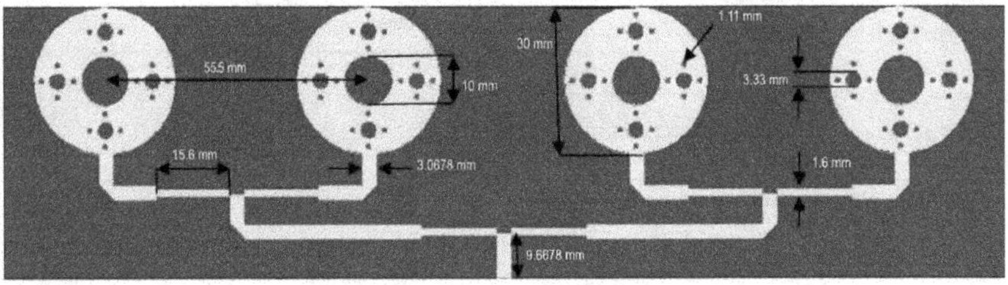

Figure 3. Design of 1×4 Linear Fractal Antenna Array

The designed antenna array is simulated with high frequency structure simulator and it is revealed that antenna resonates at several frequencies hence multiband response is achieved.

3. Results and Discussion

This section contains the output parameter for fractal antenna and antenna array configurations. From simulated results it is observed that designed antenna and arrays resonate at multiple frequencies hence output response with multiband has achieved. The performance parameters describe the efficiency of designed antenna array.

3.1 Output Parameters of Fractal Antenna

The designed circular shaped fractal antenna configuration with partial ground plane is simulated with HFSS software and performance parameters of fractal antenna are analysed. The S (1,1) parameter for 0th, 1st and 2nd iteration of designed antenna is illustrated in Figure 4. Directive gain and radiation pattern for designed antenna with partial ground plane are demonstrated in Figure 5 and Figure 6.

3.1.1 Scattering Parameter of Fractal Antenna

It is a significant parameter which defines that how much power is reflected while input power is applied. If there is no reflected power, it means there is proper impedance matching of feed line with that of patch. Impedance mismatch of patch with feedline provides the ratio of reflected power to the input power, which is known as return loss. The plot for scattering parameter s (1,1) vs. frequency for each iteration of fractal antenna has shown in Figure 4.

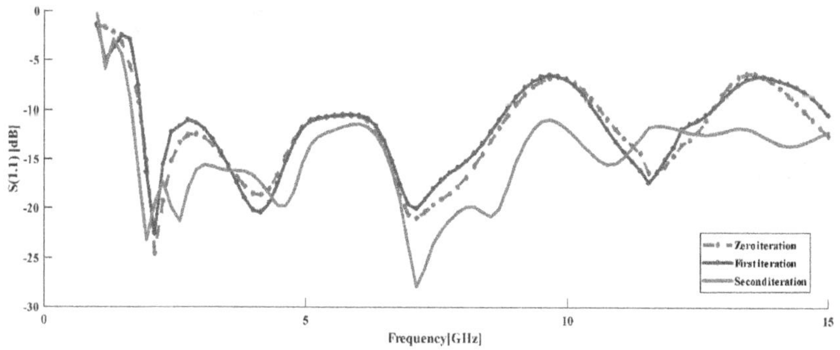

Figure 4. Scattering parameter vs. Frequency plot for Proposed Antenna

A plot in Figure 4 shows that as the number of iterations is increasing then the perimeter of designed structure is also increasing which results in shifting of frequency towards lower frequency value. As moving from basic to designed antenna structure bandwidth is also increasing. It has been analysed, for 0th iteration fractal antenna resonates at four frequencies 2.10, 4.14,7.13 and 11.57 GHz with minimum return loss of -24.509 dB. For 1st iteration minimum value of return loss -22.513 dB has obtained and antenna resonates at 2.10 GHz, 4.14 GHz, 7.13 GHz and 11.57 GHz frequencies. For desired antenna minimum scattering parameter value of -27.877 dB have been achieved and antenna resonates at 1.94 GHz, 2.57 GHz, 4.61 GHz 7.13 GHz, 8.55 GHz and 10.75 GHz frequencies respectively.

3.1.2 Directive Gain of Fractal Antenna with Partial Ground

Gain of fractal antenna is a significant parameter which defines the efficiency of an antenna. It provides the information about the ratio of radiated power in specified direction by an input antenna to the radiated power by an isotropic antenna. It basically defines that how effectively an antenna can transmit or receive the power. Gain of fractal antenna on partial ground plane has demonstrated in Figure 5.

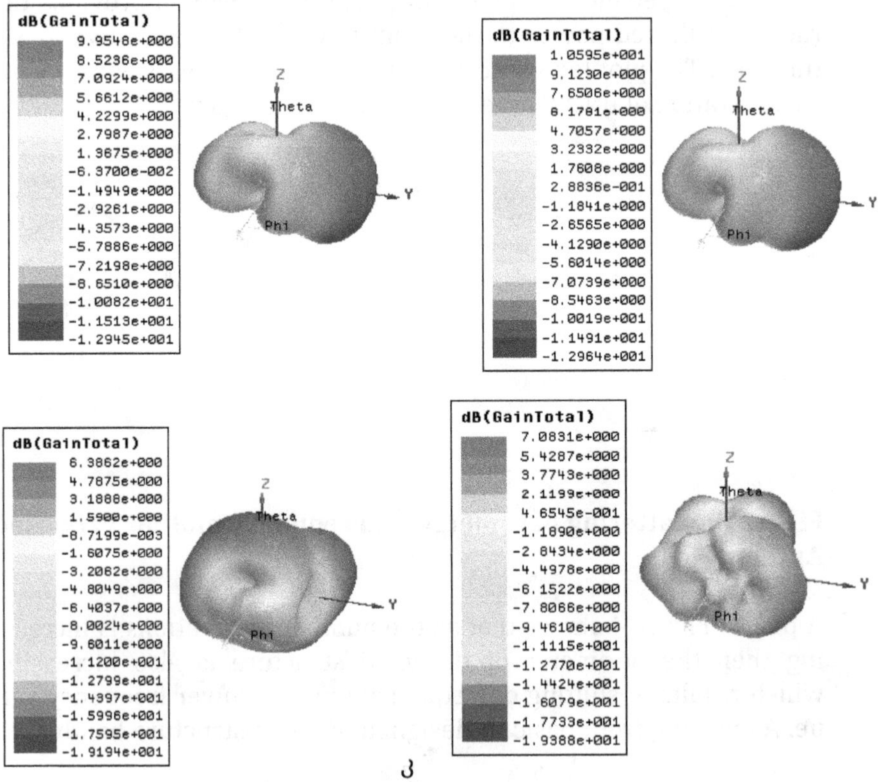

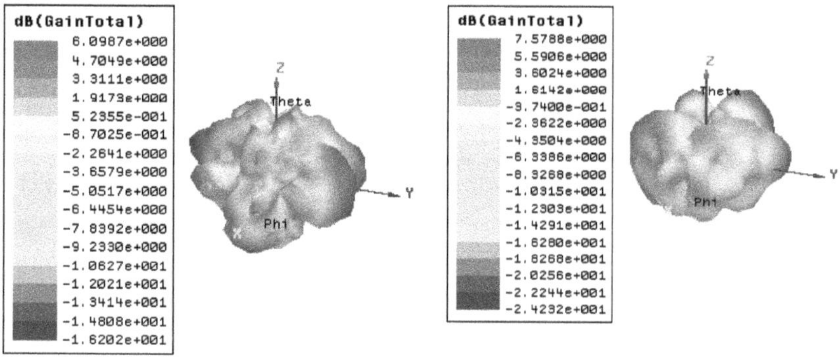

Figure 5. Gain at Frequency (top left) 1.94 GHz (top right) 2.57 GHz (middle left) 4.61 GHz (middle right) 7.13 GHz (bottom left) 8.55 GHz and (bottom right) 10.75 GHz

From the analysis of polar plots, it has been observed that maximum gain of value 10.595 dB has been achieved at 2.57 GHz frequency.

3.1.3 Radiation Pattern

It provides information about the distribution of radiation energy around an antenna.It includes both E and H plane which are orthogonal to each other.Radiation pattern at 1.94 GHz ,2.57 GHz , 4.61 GHz and 7.13 GHz, 8.55 GHz and 11.57GHz has shown in Figure 6.

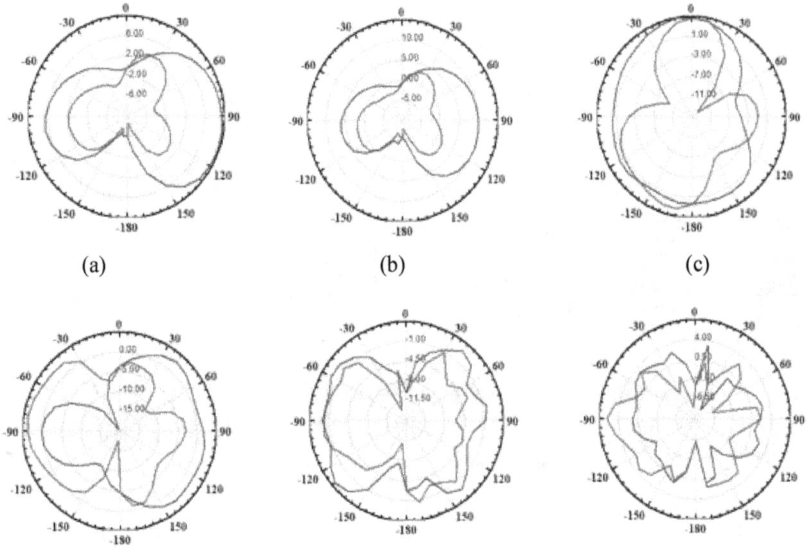

(a) (b) (c)

(d) (e) (f)

Figure 6. Radiation Pattern at Frequency (a) 1.94 GHz (b) 2.57 GHz (c) 4.61 GHz (d) 7.13 GHz (e) 8.55 GHz and (f) 10.75 GHz

Radiation pattern plot demonstrates the E and H plane which are orthogonal to each other and illustrates that bidirectional radiation patterns have achieved.

3.2 Output Parameters for Antenna Arrays

The proposed 1×2 and 1×4 antenna arrays are simulated with HFSS software and it is observed that designed arrays resonate at multiple frequencies hence output response with multiband has achieved. The simulation results of fractal antenna arrays include important parameters as Return Loss, Gain and Radiation Power Pattern which specify the effectiveness of designed antenna array. Output results for these parameters are shown in Figure 7 to 8.

3.2.1 Return Loss of 1×2 and 1×4 antenna array

Return loss is associated with the ratio of reflected back power to the input incident power. It is measured in dB and it should below -

10 dB. Further the S (1,1) parameter of 1×2 and 1×4 antenna array has been illustrated in Figure 7 and Figure 8.

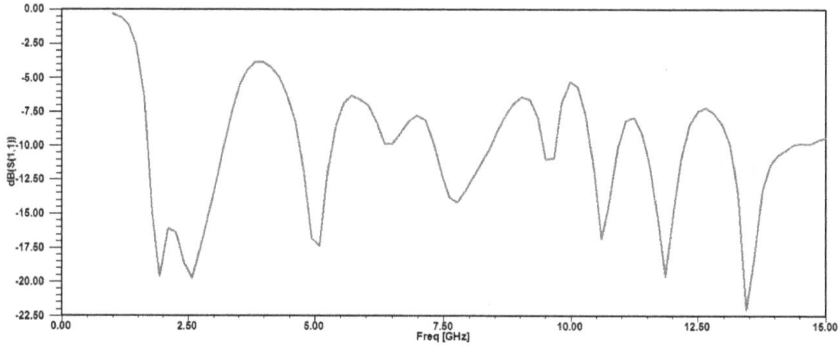

Figure 7. S (1,1) parameter vs. Frequency Plot for Two Element Array

Simulation results of designed 1×2 fractal antenna array illustrates that antenna array resonates at 1.94 GHz, 2.57 GHz, 5.089 GHz, 7.76 GHz, 9.49 GHz, 10.59 GHz, 11.85 GHz and 13.42 GHz frequencies. At each resonating frequency S (1,1) parameter values are below -10 dB which are -19.5851 dB, -19.7236 dB, -17.3472 dB, -14.1181 dB, -10.9716 dB, -16.8276 dB, -19.5735 dB, -22.0075 dB respectively.

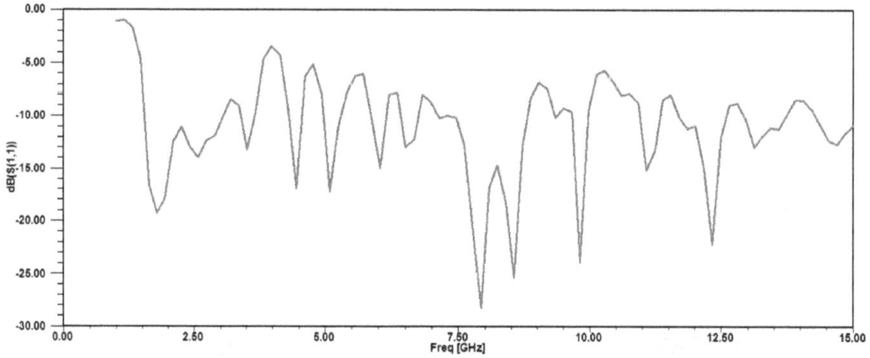

Figure 8. S (1,1) parameter vs. Frequency Plot for Four Element Array

From above plot it has been observed that antenna array resonates at numerous frequencies and return loss values are below -10 dB and gain more than 2 dB has been achieved. The values of S (1,1) parameters are -19.2210 dB, -13.9498 dB, -13.2617 dB, -16.9439 dB, -17.1947 dB, -14.9020 dB, -12.9692 dB, -28.2337 dB, -25.3634 dB, -23.8608 dB, -15.1100 dB, -22.2021 dB, -12.9597 dB and -12.7143 dB at 1.7865 GHz, 2.5730 GHz, 3.51 GHz ,4.46 GHz 5.08 GHz ,6.03 GHz, 6.50 GHz, 7.92 GHz ,8.55 GHz, 9.80 GHz ,11.06 GHz ,12.32 GHz ,13.11 GHz and 14.68 GHz respectively. Simulated results have been validated with measured results.

3.2.2 Gain and Radiation Pattern of 1×2 and 1×4 Linear Antenna Array

The Gain of fractal antenna array defines the radiation intensity of transmitted power and the power pattern plot describes the distribution of radiated power in all directions. Proposed arrays have achieved the gain more than 2 dB at each resonant frequency. Here, directive gain and power pattern are shown in Figure 9 and Figure 10 at designed frequency.

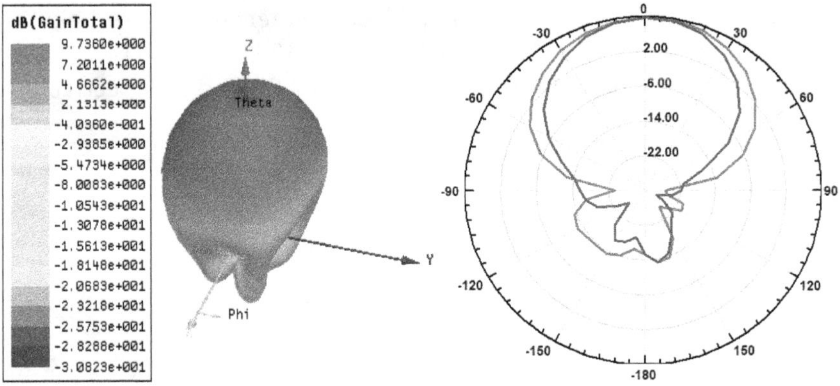

Figure 9. Gain & Power Pattern at 2.57 GHz for 1×2 Antenna Array

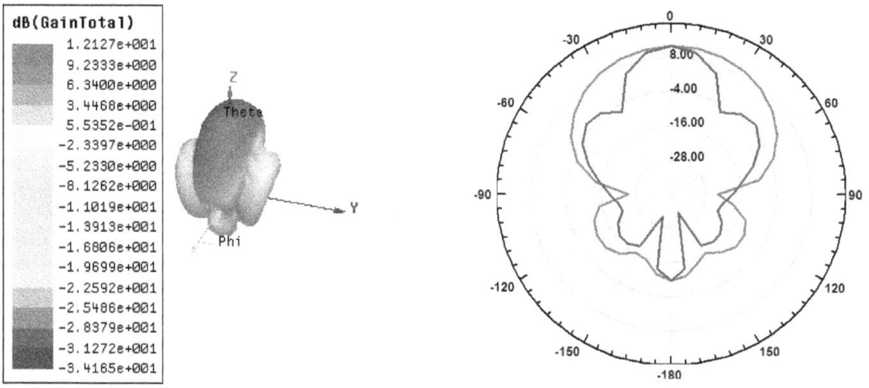

Figure 10. Gain & Power Pattern at 2.57 GHz for 1×4 Antenna Array

Simulation results of gain plot depicts that gain of 9.73 dB and 12.127 dB has achieved at 2.57 GHz resonating frequency for linear fractal antenna array with two and four elements respectively. Radiation pattern shows the energy distribution in both directions. Hence bidirectional radiation pattern has been obtained.
Numerical values of performance parameters of antenna arrays at each resonating frequency are summarized in Table 3.

Table 3. Numerical values of performance parameters of Antenna Arrays

Antenna Array	Resonant Frequency [GHz]	S (1,1) [dB]	Gain [dB]
1×2	1.94, 2.57, 5.08, 7.76, 9.49, 1059, 11.85, 13.42	-19.5851, -19.7236, -17.3472, -14.1181, -10.9716, -16.8276, -19.5735, -22.0075	9.19, 9.73, 7.97, 8.44, 6.79, 7.07, 9.50, 6.85
1×4	1.78, 2.57, 3.51, 4.46, 5.08, 6.03, 6.50, 7.92, 8.55, 9.80, 11.06, 12.32, 13.11, 14.68	-19.2210, -13.9498, -13.2617, -16.9439, -17.1947,	10.383, 12.127, 11.571, 9.446, 9.166, 6.671, 6.998, 7.634, 8.247, 5.701, 11.306, 7.439, 7.896, 8.280

		-14.9020, -12.9692, -28.2337, -25.3634, -23.8608, -15.1100, -22.2021, -12.9597, -12.7143	

In Table 3 antenna array's performance parameters are analysed and numerical values of return loss and gain at various resonating frequencies are presented. it has been observed that minimum return loss of -22.0075 dB and maximum gain of 9.73 dB has been achieved for 1×2 antenna array and it has been revealed that for 1×4 fractal antenna array, minimum return loss of -28.2337 dB and 12.127 dB gain has been achieved.

4. Comparison of performance parameters of Fractal Antenna Array

The linear fractal antenna array based on circular shape fractals is designed along with 2 and 4 number of patch elements and simulated with HFSS software. Patch elements have been arranged in a linear mode. As gain parameter determines the efficiency of antenna array so, peak gain of antenna arrays has been compared and analysed in graphical form. Peak gain vs. frequency plot for 1×2 and 1×4 antenna array is shown in Figure 11.

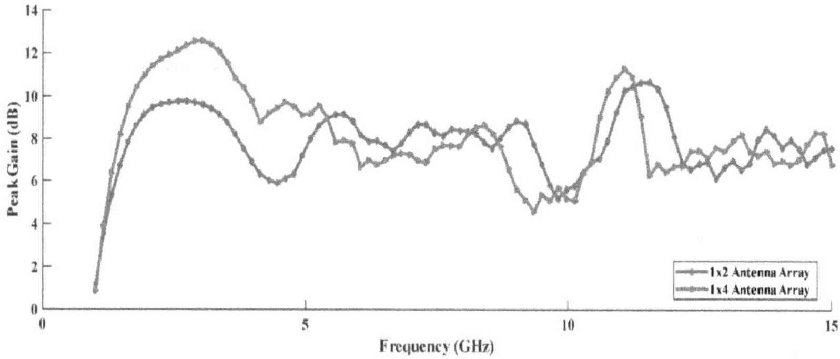

Figure 11. Peak Gain vs. Frequency for 1×2 and 1×4 Antenna Array

Figure 11 depicts the plot of parameter gain for 1 ×2 and 1× 4 antenna arrays. Numerical values of performance parameters are presented in Table 4.

Table 4. Comparison of Performance parameters of Antenna Arrays

Antenna Array		No. of elements	Peak Gain (dB)	Return Loss (dB)
Linear	**1 × 2**	Two	10.67	-22.0075
	1 × 4	Four	12.59	-28.2337

It has been concluded that as the number of patch elements is increasing frequency is shifted towards the lower side and an increase in gain and improved return loss has been achieved for the linear antenna array.

Conclusions

Antenna array configurations are proposed with circular shaped fractals on partial ground plane. For 0th and 1st iteration of designed single element antenna, two wide bands of frequencies 7.12 GHz and 2.13 GHz, 7.01 GHz and 2.21 GHz are obtained. For the last iteration, a wide band of 13.345 GHz frequency has been achieved. But antenna array with two and four elements resonates at multiple frequencies and it has been resolved that an increase in the number of patch elements leads to minimum return loss -28.2337 dB and higher gain 12.59 dB. It covers L, S, C and X and Ku band of frequencies which can be used for WiMAX, WiFi devices and satellite applications.

References:

[1] Bisht, S., Saini S., Bhaskar V. P. (2014) Study the various Feeding Techniques of Microstrip Antenna Design and Simulation Using CST Microwave Studio. *International Journal of Emerging Technology and Advanced Engineering.* **4**(9).
[2] Pozar D. M. (1996) A Review of Aperture Coupled Microstrip Antennas History, Operation, Development and Applications. *IEEE Letters,* **35**, 1-38.
[3] Dang, L., Lei Z. Y., Xie Y., Ning G. L., Fan J. (2010) A compact microstrip slot triple band antenna for WLAN/WiMAX applications. *IEEE Antennas Wireless Propagation Letter,* **9**, 1178-1181.
[4] Yang X, W. Hu, Y. Yin, P. Fei. (2011) Compact triband square slot antenna with symmetrical L strips for WLAN/WiMAX applications. *IEEE Antennas Wireless Propagation Letter,* **10**, 462-465.
[5] Kumar A., A.K. Gautam, B.K. Kanaujia (2013) An annular ring slot antenna for CP operation. *Microwave opt Techno Lett*er, **55**(6), 1418-22.
[6] Khandelwal M.K., B.K. Kanaujia, S. Dwari, S. Kumar, A. Gautam (2014) Analysis and design of wide band microstrip line fed antenna with defected ground structure for Ku band applications. *AEU – International Journal of Electronics and Communications,* **68**(10), 951-957.
[7] Khandelwal M.K., B. K. Kanaujia, S. Dwari, S. Kumar, A. Gautam (2015) Analysis and design of dual band compact stacked microstrip patch antenna with defected ground structure for WLAN/WiMAX applications. *AEU–International Journal of Electronics and Communications,* **69(**1), 39-47.
[8] Arya S., S. Khan, C. Shan, P. K. Lehana (2012) Design of a Microstrip Patch Antenna for Mobile Wireless Communication Systems. *Journal of Computational Intelligence and Electronic Systems,* **1**(2), 178-182.
[9] Arora R., A. Kumar, S. Khan, S. Arya (20114) Design Analysis and Comparison of HE shaped and E Shaped Microstrip Patch Antennas. *International Journal on Communications Antenna and Propagation,* **4**(1), 27-31.

[10] Wong, K. L. (2017) *Compact and Broadband Microstrip Antennas*, John Wiley & Sons.
[11] Puente, C. J. Romeu, R. Pous, J. Ramis, and A. Hijazo (1998) Small but long Koach fractal monopole. *IEEE Electronic Letter*, **34**, 9-10.
[12] Anguera, J. C. Puente, C. Borja, and J. Romeu (2000) Miniature wideband stacked patch antenna based on the Sierpinski fractal geometry. *IEEE Antennas and Propagation International Symposium*, Salt Lake City, UT, **3**, 1700-1703.
[13] Anguera, J. C. Puente, C. Borja, R. Montero, and J. Soler (2001) Small and high-directivity bow-tie patch antenna based on the Sierpinski fractal. *Microwave Optical Technology Letter*, **31**, 239-241.
[14] Anguera, J. E. Martinez, C. Puente, C. Borja, and J. Soler (2004) Broadband dual-frequency microstrip patch antenna with modified Sierpinski fractal geometry. *IEEE Transactions on Antenna and Wave Propagation*, **52**(1), 66-73.
[15] Bisht, N. and P. Kumar (2011) A dual-band fractal circular microstrip patch antenna for C-band applications. *PIERS Proceedings*, Suzhou, China, 852-855.
[16] Bangi I.S, J.S. Sivia (20118) A Compact Hybrid Fractal Antenna using Koch and Minkowski Curves. *9th Annual Information Technology, Electronics and Mobile Communication Conference (IEMCON).*
[17] Desai, Arpan, Trushit K. Upadhyaya, Riki kumar Hasmukh bhai Patel, Sagar Bhatt, and Parthesh Mankodi (2018) Wideband high gain fractal antenna for wireless applications. *Progress in Electromagnetics Research*, **74**, 125-130.
[18] Bhatt, Sagar, Parthesh Mankodi, Arpan Desai, and Riki Patel (2017) Analysis of ultra-wideband fractal antenna designs and their applications for wireless communication A survey. *International Conference on Inventive Systems and Control (ICISC)*, pp. 1-6.
[19] Antoniades ,A.M., G. V. Eleftheriades (2008) A compact multiband monopole antenna with a defected ground plane. *IEEE Antennas Wireless Propagation Letter*, pp. 652-655.
[20] K. H. Chiang, K. W. Tam (2008) Microstrip monopole antenna with enhanced bandwidth using defected ground structure. *IEEE Antennas Wireless Propagation Letter*, pp. 532-535.
[21] Balanis ,C.A.(2005)*Antenna theory: Analysis and Design*, 3rd edition, Wiley.
[22] Lakshmana Kumar V.N., M. Satyanarayanana, P.V. Sridevi, M.S. Parakash (2014) Microstrip fractal linear array for multiband appli-

cations. *IEEE International Conference on Advanced Communication Control and Computing Technologies*, pp. 1818-1822.
[23] Bernety H.M., R. Gholami, Bijan and M. Rastamian (2013) Linear antenna array design for UWB radar. *IEEE Radar Conference.*
[24] Barison A., P. Deo and D.M.Syahkal (2014) A switched beam 60 GHz 2 x 2 planar antenna array. *IEEE 8th European Conference on Antenna and Propagation (EuCAP)* pp. 1000-1002.

Chapter 2

Advanced Research Perspectives in Modelling of Diseases using R

Dheva Rajan. S
Mathematics Section, Department of Information Technology, University of Technology and Applied Sciences-AlMusannah, Sultanate of Oman

1. Introduction

According to Britannica [1], a disease is "any harmful deviation from the normal structural or functional state of an organism, generally associated with certain signs and symptoms and differing in nature from physical injury". In Biology, the knowledge mining of illness is termed "Pathology", which includes the study of Etiology (classification of disease or cause). [2], [3] [4, p. 9], [5], [6] provided various types and classifications of diseases. The flowchart in figure 1, created using https://app.diagrams.net/, gives us the consolidated classification of diseases. The flow of this chapter is highlighted with a larger font size in figure 1. Here, vector-borne disease (VBD), called Dengue Fever (DF), is taken as a counter-example to mathematical modelling. In this chapter, equal priorities are given to both mathematical modellings and solutions through programs. DF is caused by the mosquito family "Ades", especially by two specious Ades Aegypti (AA), Ades Polynesiensis and Ades Albopictus. Their eggs are able to be sustainable even in the nonexistence of water for several months' Adult males, and females both eat nectar and flower juices. When a healthy mosquito bites a human, that become infected and if that infected mosquito biting a human, human become infected. There is no vaccine for DF, but a vaccine called Dengvaxia® is under evaluation. For more details on dengue vaccine one can refer the Elsevier article extracted from WHO [7]. While most mosquitos bite at night, dawn, sunset, AAM bites both inside and outdoors during the daytime. If an AAM bites an infective person, it can spread the disease after seven days. However, modelling of mosquito-borne disease starts with Malaria. The first dengue epidemic in India was identified in Chennai, (formerly Madras) Tamil Nadu. The first proven epidemic of DF happened at Kolkata (formerly Calcutta) between 1963 and 1964 [8]. There are two types of DF. One is DF, and another is Sever Dengue (Dengue Hemmograric Fever (DHF)). The Virus that can yield Dengue is called Dengue virus, notated as DenV. There are four strains of DenV, say, DEN1, DEN2, DEN3 and DEN4. Different serotypes or a mixture of strains are found in the world at various places. A person infected by one strain has the chance of getting infected with other strains. The fatality rate of DHF is less than 1%. The consolatory remark is DF is mild. DF can occur without any symptoms at all, said to be asymptotic. We will be using this terminology in the latter part of this chapter to explain the analysis. [9] Massed et al., at 2011 indicated that around 80 percent of are asymptotic when infected by DenV. Female mosquito will suck about five millionths (or 0.000005) of a litre of blood in a single serving and they can drink 3x their weight in blood.

The foremost mathematical modelling of epidemics was in 1766 by Daniel Bernoulli to analyze the morality of smallpox [10]. In 1772, Lambert developed the work of Bernoulli with age structuring [11]. Since such investigations were not lined up correctly and scattered, a systematic concept was published by Ross in 1911 for Malaria; usually mentioned by many as the first work of mathematical modelling of epidemiology [12]. Though [12] was discovered in 1902, he has not published/open that till 1909. [13] and [14] proposed the developed work of [12] with the compartmental idea. [12] addressed prior modelling, whereas a detailed review and criticism of this model have been made by [15]. The mathematical readers of this chapter do not want to worry about a few biological terms used regret about that, but that become essential and easy to refer to/understand too. Hence, under the non-communicable diseases, Dengue has been considered a counter-example to develop the high-level model, whereas the lower levels were mentioned with various developments in common to other diseases. It is not execration to say mathematical models (MM) rule the world. There are many illustrations like weather forecasting, rain forecasting, finance, national census related forecasting, and umpteen of science and engineering that elucidates mathematics's applications.

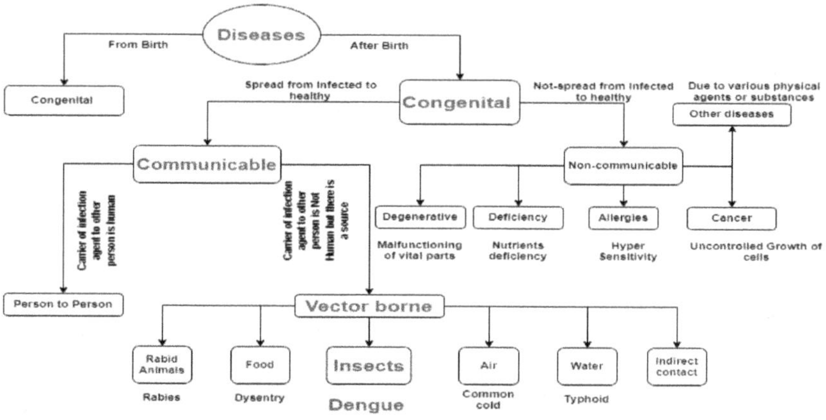

Figure 1: Classification of diseases

For instance, [16] has done work on chemical applications of graph theory, [17] on A study on the graph-theoretical approaches on molecular biology, [18] on a survey of the graph-theoretical properties of chemical molecules, [19] on A survey of Nonparametric statistical linkage for dichotomous traits. Even common cold is an infectious disease through the medium, say air. A probability transition matrix is used to analyze the probability of infection [20]. A mathematical model for natural cooling of a cup of tea was proposed by [21], and that becomes the fundamental model for developing further for infectious disease modelling through ordinary differential equations (DE will denote ODE, the differential equation); in ODE, the unknown variable is a function of a sole independent variable. Many kinds of research are going on with Dengue still; the objective of this chapter is to get into the sequential development of mathematical modelling of "Dengue" and proposed developments. The DF is a VBD, not with human to human (HH) spreadable pattern.

2. History of Dengue and the trend in India

For instance, the dengue cases and deaths in India are given in Figures 2 and 3.

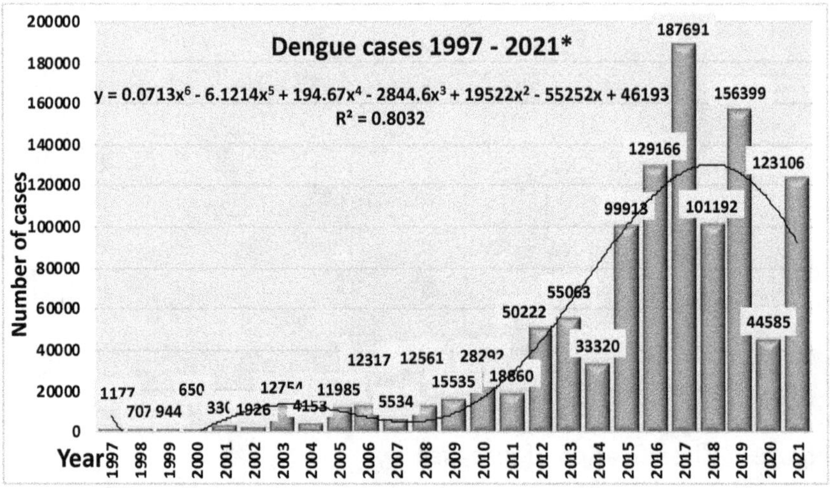

Figure 2: Dengue cases in India (Source: https://nvbdcp.gov.in/)

A sixth-degree polynomial provides a reasonable approximation for the dengue cases and deaths than the other basic models. But still, the coefficient of determination is 0.602 for fatalities; on the other hand, it holds a value of 0.8032 for DF cases. The economic loss due to DF is more worldwide. Many have given the economic view of DF. For instance, [22] did for the US in 2011, [23] worked on the economic impact of DF in Singapore between 2010 and 2020. [24] analyzed for India in 2019, [25] investigated the productivity cost of DF in the Asian region. [26] did a systematic global analysis in 2016 and [27] in 2020.

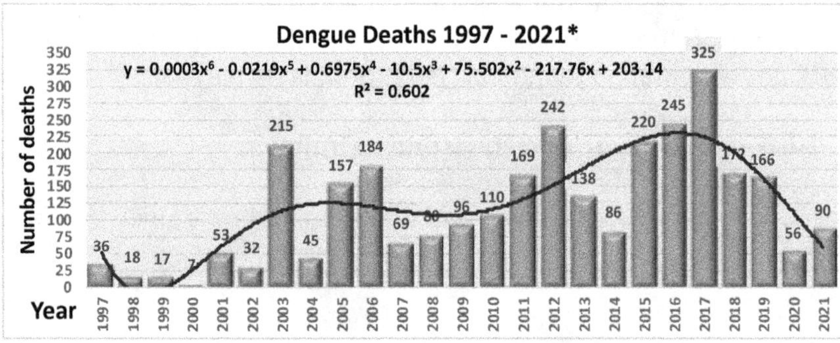

Figure 3: Dengue deaths in India (Source: https://nvbdcp.gov.in/)

3. Modelling

In the early days, people thought that the only possible way was to eradicate the vector from this world, and no other routes were likely to eliminate the infectious VBD.

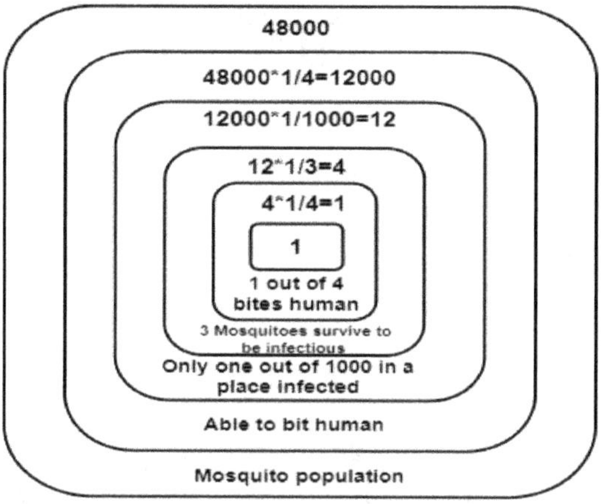

Figure 4: Ross mosquito theorem

Ross [9], the first person, proved that was a misconception, and it is enough to make the susceptible (A susceptible individual is a fellow of a cluster in a population is having the possibility of contracting an infection) safe by using various factors. Ross [9] proposed the theorem known as the "Ross Mosquito theorem". Ross's idea is a simplified calculation in figure 4, drawn using https://app.diagrams.net/. An infected individual is a fellow of a cluster from a susceptible population (usually) confirmed the infection. Let g be the fraction of the infected individuals. Then $(1-g)$ become the fraction of the susceptible individuals. Let t be the elapsed time. So here,

g=1/6 and $(1-g)$ =1-(1/6) =5/6

Now, the rate of spread concerning time is proportional to the rate we found. Hence,

$\frac{dg}{dt}\alpha g(1-g) \Rightarrow \frac{dg}{dt} = \beta g(1-g)$, β is the proportionality constant

Now, one can solve the above equation.

By assuming β = *1*, we can get the following table of values.

g	0	0.1	0.2	0.3	0.4	0.5	0.6	0.7	0.8	0.9	1
(1-g)	1	0.9	0.8	0.7	0.6	0.5	0.4	0.3	0.2	0.1	0
Rate	0	0.09	0.16	0.21	0.24	0.25	0.24	0.21	0.16	0.09	0

To calculate the susceptible and infected ratio, consider the following 6x6 matrix of population, at which we assume that the highlighted cells (ranging from 1 to 6) are infected, and the rest of the people are susceptible.

Table 1: Susceptible, infective assumption

1	7	8	9	10	11
12	13	2	14	15	16
17	18	19	20	3	21
22	4	23	24	25	26
27	28	29	30	31	32
33	34	5	35	6	36

Hence, the susceptible ratio is *1-g = 30/36 =5/6* with *g = 6/36 = 1/6.*

$\frac{dg}{dt} = \beta g(1-g)$ where β is a positive constant. On solving, one can get, $g = \frac{1}{1-\left(1-\frac{1}{a}\right)e^{-\beta t}}, t=0, g=0$

The "Extrinsic Incubation Period" (EIP) is defined as "The time it takes from ingesting the virus into the actual transmission to a new host". At ambient temperature between 25 to 28 degrees Celsius, the EIP for dengue is between 8 to 12 days [28], [29],and [30]. Before the symptoms occur in the human body, the human to mosquito (H-M) transmission can happen [29], [31]. Even after two days of recovery from fever, such H-M transfer can happen. [32]. [33, pp. 2018–2022] provided forecasting for DF till 2022, and the parameter values almost coincide with the specified one.

3.1 Calculation of Growth rate

Assume that one infected person initially passes the infection to two persons.

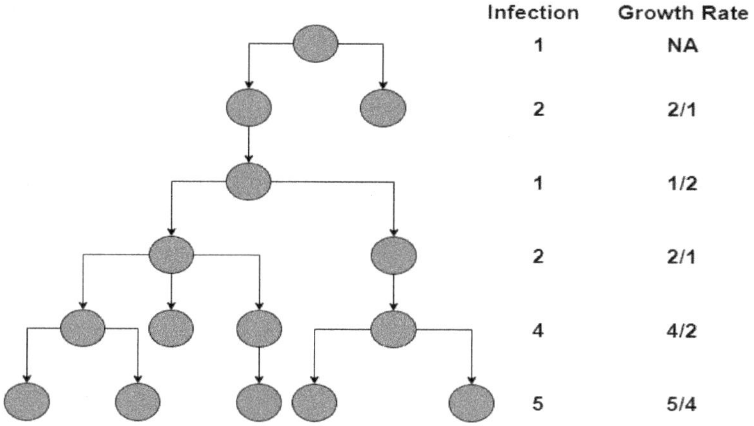

Figure 5: Infection pattern and Growth rate

The assumption is two persons will pass the infection to the other two, and it goes on. So, the sequence of number become, 1,2,4,6,16,32,64,128,256,512,1024,2048, ... But in reality, not all persons pass the infections to others due to quarantine, medications, and other factors. Also, one infected person can spread the infection to more than one person. That scenario, along with the growth rate, is explained in figure 5, created using https://app.diagrams.net/.

4. Model development, Solution and proposed future developments

Assume the total population of any specious (here say human) at any time *t* is N(*t*)=N. The proportion of variation is a DE for N. The function of *t* notation (*t*) will be removed when creating the equation for the sake of simplicity. Hence the model by Ross [9] can be explained as follows:

$$\frac{dN}{dt} = births - deaths; \quad \frac{dN}{dt} = \beta N - dN; \quad N(t) = N_0 e^{(\beta - d)t}$$

Transmission coefficient, say β = contact rate times transmission probability. In this connection, one should understand that there are many new generation calculations for transmission coefficient, β [34] and [35]. The first law of Kirchoff law of current states, "the current flowing into a node (or a junction) must be equal to the current flowing out of it". By using the same token, however, the total population will never alter the number of stages one is assuming. The initial set of dependent variables counts the number of members in a population in every cluster as a function of time. Let S = S(t) denote susceptible count; I = I(t) represent the number of diseased, called model states. The basic idea of developing the SI model for the disease is as follows.

Change in the susceptible =

$$\frac{dS}{dt} = -\beta * Current\ Susceptible\ population * Infected\ population$$

Change in the infected =

$$\frac{dI}{dt} = Growth\ throuth\ new\ infection - loss\ due\ to\ recovery$$

$$\frac{dI}{dt} = \beta * Current\ Susceptible\ population * Infected\ population - \gamma * Infected\ population$$

There should be at least one infection at any time 't' to occur the epidemic. Hence, $\frac{dI}{dt} > 0$ this means, $S > \frac{\gamma}{\beta}$, at the *t=0*, in this case, S at *t=0* is *N*, the whole population. Therefore, the basic reproductive number R0 = R at t=0 = N* $\beta * \tau$ is the force of infection. As a separate topic is dedicated to R0 at the later part of this chapter, we stop finding the value. The development happens when the people think about the recovered individuals. Unless the recovered individuals are included, there will be a loss of accuracy in the prediction. Hence, the recovered individual is a fellow of a cluster in an infected population recovered from the infection. Let R = R(t) denotes the count of recovered individuals. Here, S, I and, R is known as "States", explained in figure 6 and S+I+R=N. As one should deal with large numbers to simplify the process, fractional quantities are used. The fractional quantities will be obtained by dividing all the state quantity numbers by the total population.

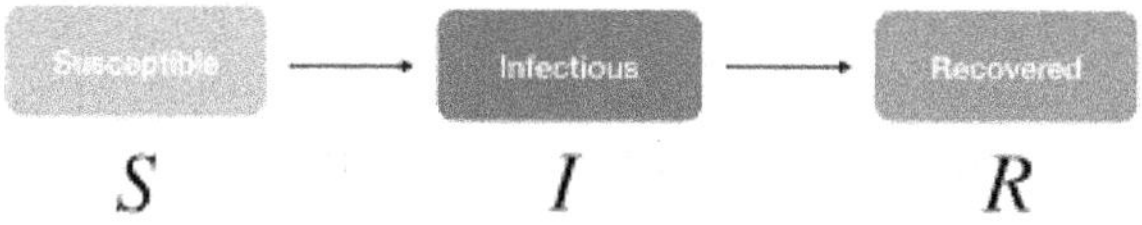

Figure 6: SIR model

Hence, Let s=s(t) = S(t)/N denotes the fraction of susceptible, i=i(t)= I(t)/N denotes the fraction of infected population, and, r = r(t) = R(t)/N denotes the fraction of recovered. The origin of such models started with [13]. Earlier it was with only S and I; later, in 1927, [13] proposed a set of equations explaining S, I and R. Here, the SIR model is defined as a recursion relation (RR) for better performance.

$\frac{dS}{dt} = -\beta S(t)I(t)$ and the fractional equation becomes $\frac{ds}{dt} = -\beta s(t)i(t)$, where β is the infection rate. The negative sign here is used since that quantity will be removed from S and moving towards I.

$\frac{dR}{dt} = kI(t)$ and $\frac{dr}{dt} = ki(t)$ where k is the recovery rate. So,

$\frac{dI}{dt} = \beta S(t)I(t) - kI(t)$ and $\frac{di}{dt} = \beta s(t)i(t) - ki(t)$.

These three equations constitute a model for the spread of the disease, called the SIR model. People are unaware of β and k values, but they can be estimated. Note that the infection rate β is a fixed number of contacts per day sufficient to spread the disease by an infected individual, and the recovery rate k is a recovered fraction from infection during any given day. Here, it is to be noted that,

$$\frac{ds}{dt} + \frac{di}{dt} + \frac{dr}{dt} = 0.$$

Let's start doing it manually through an illustration. After this, all the equations are solved directly through R programming. Consider an arbitrary example of the SIR model. Suppose for a disease the population at the time of starting the disease is 7900000 the number of infected individuals is ten, and there is no recovery. Hence, the initial condition of SIR model is S (0) =7900000, I (0) =10, and R (0) =0. As one must deal with large quantities, using the fractional equation is always better. Hence, s (0) =S (0)/N, i(0) =I(0)/N, r(0)=R(0)/N. So, s(0)=1, i(0)=1.27x10^{-6} and r(0)=0.

The equation now is,

$$\frac{ds}{dt} = -\beta s(t)i(t) \qquad ,s(0)=1; \qquad \frac{di}{dt} = \beta s(t)i(t) - ki(t) \qquad ,i(0)=1.27\times10^{-6}$$

$$\frac{dr}{dt} = ki(t) \qquad ,r(0)=0$$

The solution of the above equation set is given in figure 7 drawn using R programming. The code is given below.

```
require(deSolve); params <- c(beta=400, gamma=365/13) ;
times <- seq(from=0,to=60/365,by=1/365/4) ; xstart <-
c(S=0.999,I=0.001,R=0.000)
closed.sir.model <- function (t, x, params) { S <- x[1] ; I <- x[2] ; R <-
x[3] ; beta <- params["beta"] ; gamma <- params["gamma"] ; dSdt
<- -beta*S*I ; dIdt <- beta*S*I-gamma*I ;
  dRdt <- gamma*I ; dxdt <- c(dSdt,dIdt,dRdt) ; list(dxdt) }
out <- as.data.frame( ode( func=closed.sir.model, y=xstart,
times=times, parms=params ) )
plot(I~time,data=out,type='l',col="blue",lwd=3,xlab="time t", ylab =
"No.of.persons",box.lwd=3)
lines(S~time,data=out,type='l',col="red",lwd=3)
lines(R~time,data=out,type='l',col="darkgreen",lwd=3)
title(main = "Plot of SIR model" ); box(box.lwd=3)
legend("topright", legend=c("S", "I","R"),col=c("red",
"blue","darkgreen"), lty=1, cex=0.8,
     box.lty=1, box.lwd=3, box.col="brown")
```

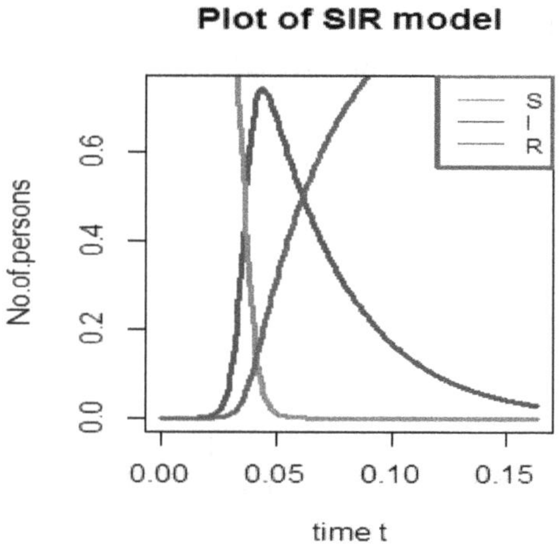

Figure 7: Solution of SIR model

R programming is a valuable tool for data analysis and visualization. [36] It is highly applicable in data science data analytics, and many academics are using this program in their statistics courses. Few people attempted to escalate this programming to solve mathematical ideas, too. "deSolve" is among various packages of Differential equation solving. There are many packages available in R programming. As this chapter consists of only ODE modelling, it gives insight into the packages in ODE. The first package is "odesolve", which has two integration methods. "odesove" has been replaced by "deSolve" as it has many other solvers too, like ODE, DDE, DAE, and PDE [37]. It can solve stiff and non-stiff problems also. In Julia programming, there is a package "DifferentialEquations.jl", and in R, the excellent alternative to "DiffeentialEquations.jl" is "diffeqr". The models are compatible for both R and Julia if the Julia language is also installed along with R in the system [38]. "odeintr" is the integrated package that can compile packages from Rcpp-C++ODE solvers and "boost odeint" [39]. Runge Kutta (RK) method is the most widely used one by the researchers to solve ODE, such adaptive RK method solvers ode23, ode3s, ode45, Burlisch-Stoer method and "pracma" package consists of all the methods and the integration of all those packages [40]. For physics and engineering problem solving, "rODE" [41], for performing Cvode function "sundialr" in which the program should be written in R or Rcpp function [42], for shorthand prescription dosing, "mrgsolve" [43], and nonlinear mixed effects of modelling, "RxODE" [44].

Coming back to the solution of the SIR problem discussed above, the same set of equations is explored with changing initial conditions, and by using the below program gives us figure 8.

```
library(deSolve)
    sir <- function(time, state, parameters) {
    with(as.list(c(state, parameters)), {
    dS <- -beta * S * I ;    dI <- beta * S * I - gamma * I ;    dR <-
    gamma * I ;
    return(list(c(dS, dI, dR))) })}
    init <- c(S = 1-1e-6, I = 1e-6, R = 0.0)
    parameters <- c(beta = 1.4247, gamma = 0.14286) ; times <-
    seq(0, 70, by = 1)
    out <- ode(y = init, times = times, func = sir, parms = parame-
    ters) ;  out <- as.data.frame(out)
    out$time <- NULL
    matplot(x = times, y = out, type ="l",bty = "l",lty = 1, xlab =
    "Time (t)", ylab = "S, I and R", main = "SIR Mod-
    el",col.main="red",col.axis="red", lwd = 2, col = 1:3)
    matplot(x = times, y = out, type="o", xlab = "Time (t)", ylab = "S, I
    and R",        main = "SIR Model",col.main="red", lwd = 2,lty=1:3,
    pch=1:5, col = 1:5)
    axis(1,col.axis="blue",font = 2); axis(2,col.axis="blue",font = 2)
    legend(35, 0.8, c("Susceptible", "Infected", "Recovered"), pch =
    1:5, col = 1:3, lty =1:5,cex=1,text.font = 1,y.intersp = 0.7); box(lty
    =14,col='red')
```

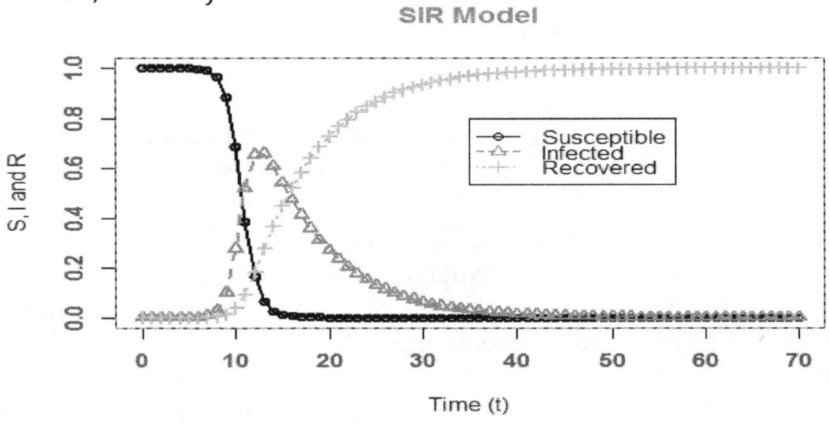

Figure 8: Solution of SIR model with varying initial conditions

The modelling equation becomes that no one is added to the susceptible group since we ignore births and migration. Hence, the above model can be developed by adding the parameters for birth and migration. Sorry for the dual usage for the total N and TP population; it eliminates the practical difficulties.

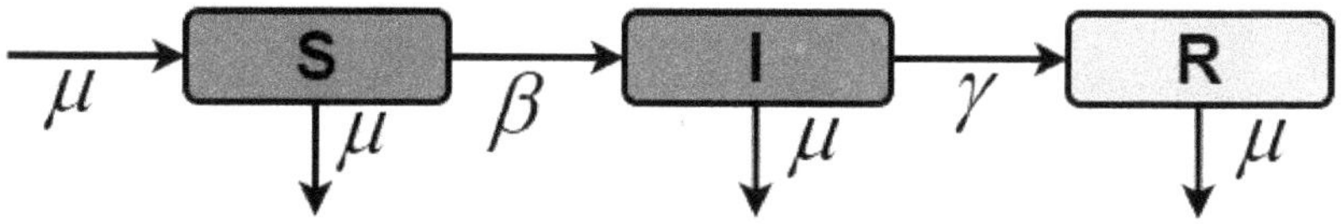

Figure 9: SIR model birth and death

Figure 9 is the state diagram used to propose the below model with migration, birth and death parameters.

$$\frac{d(TP)}{dt} = births\ -\ deaths\ +\ migration;$$

$$\frac{dS}{dt} = births\ -\ \beta SI\ -\ death * S;$$

$$\frac{dI}{dt} = \beta SI - \gamma I - \ death * I\ ; \qquad \frac{dI}{dt} = \gamma I - \ death * R$$

where γ is the rate of recovery.

The solution is given in figure with initial conditions is given in figure 10. On using "deSolve" package with matplot command, the predicted values are given as follows: beta =0.100, recovery=0.005, death =0.001, birth =0.001. it can be predicted that the vaccination is also here, but as a separate section is devoted, it is left here.

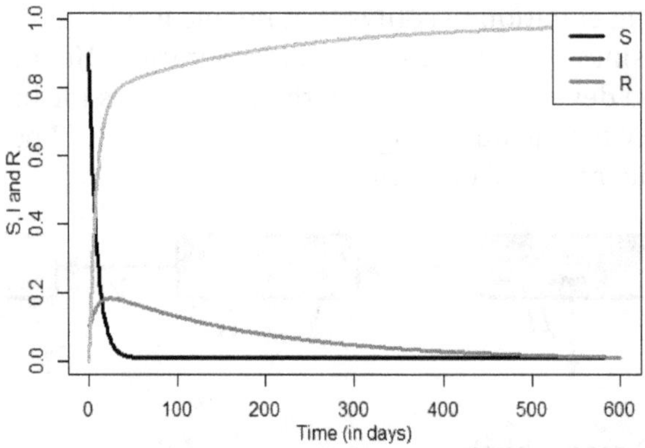

Figure 10: Solution of SIR model birth and death with varying initial conditions

Now, one may have the question, how to get the infection rate and recovery rate? Simply the recovery rate can be defined as "the inverse of the average infectious period", whereas the infectious period can be defined as the "period from the first obvious appearance of an illness to the fullest recovery from the disease". One can go through the articles by [45], [46], and [47] to get more insight into recovery rates.

5. Analysis and Further Developments

5.1 The Basic Reproduction number R0

The R0 for a disease plays a critical part in assessing the disperse of that sickness. λ, the infection force, can be defined as the "Per capita rate at which susceptible individuals contract the infection". The new virulent are created at a rate λ times X, where X = n*r times of susceptible. The basic reproduction number (R0), articulated as "R nought," is the furthermost essential parameter in assessing the contagiousness of the disease. There are many different ways to calculate the reductive number R0; we mentioned the basic one (that's why it is called Basic R0). In 1990, [48] published an article in computation R0 in a heterogeneous mixture of populations. For instance, [49] proposed a technique to find R0 that consists of many probabilistic parameters. Now, on average, for an infected person, the infectious period is $1/\gamma$ days, and every day this will create β susceptible individuals. Hence, the newly infected individual generates throughout the contagious period is $\beta * 1/\gamma = \beta / \gamma$. (Remember the R0 formula we have given at SI model). The author of this chapter strongly recommends the readers to go through the article about the journey of R0 from 1950 onwards till COVID 19. [50, p. 19]. [51] pointed many deviations between the theoretical calculations and definition of R0 and the actual calculation through the data obtained. Figure 11 illustrates the sequential development of the disease if the R0 value is 3.

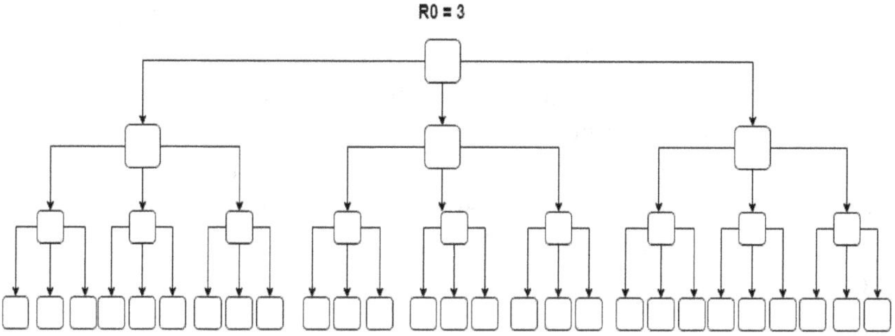

Figure 11: R0=3 if each individual produces 3 infected individuals

This R0 value may vary for different diseases. Every disease has its own R0 depending on the severity and the spreading capabilities.

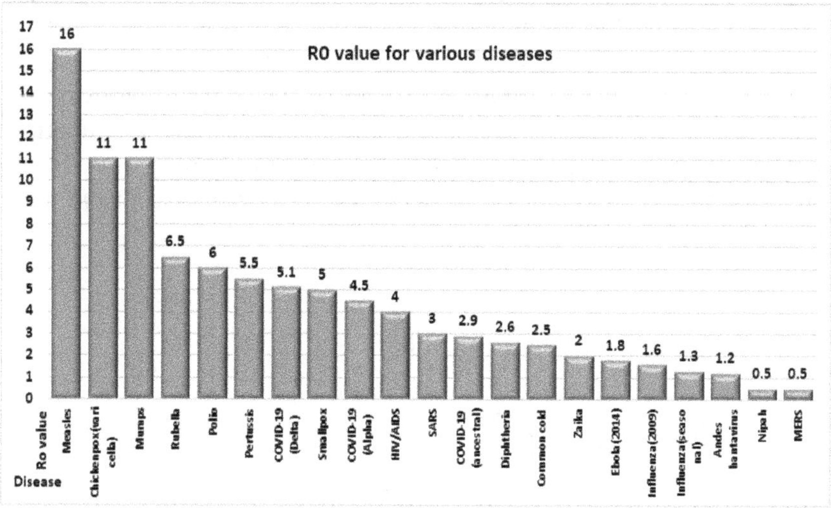

Figure 12: R0 value for various diseases Source: D. Adam [52]

This R0 can vary with time, the demographical level, the models used. Also, the immunity level of the people dramatically affects the value of R0. Figure 12 illustrates the importance of R0 for various diseases. In 2019, [53]published an article in CDC regarding the computational Complication of the R0 by discussing multiple factors and multiple forms of R0. In 2020, Janssen [54] wrote an article mentioning the economic value of R0 that elucidates the importance of new generation R0. [55] contributed a calculator to find basic parameters of disease. An exciting thing to notice with [55]; this article has 45 author-contributors. Hence, though the publishers requested the chapter contributors not to provide et al. in the bibliography, for [55] alone, et al. is provided, and the same followed for [56] as it has 17 authors. [57] discovered epidemic calculators at 2020 for covid19, [58], [59] found such epidemic calculator for any disease, [60] discovered only for mortality rates of coivd19 and [61] at 2021 for specific parameters of infectious diseases.

5.2 Effective Reproductive Number and immunity inclusion

The R0 alone is insufficient to assess the spread, but it is the essential quantity. The Effective Reproduction Number R. This can be demarcated as the product of R0 and the fraction of the S population. So, R = Fraction(S)*R0. Let's explore this idea through a simple example. For measles, assume that the R0 is 14 and 50% of the population has already been immunized; what could be the effective reproduction rate? The answer is 14 x 0.5 = 7. It means that every infected individual would produce seven cases of secondary infection. Let's continue with the new herd immunity threshold (ν). v is the fraction of the population that necessities to be insusceptible for a disease to be restricted.

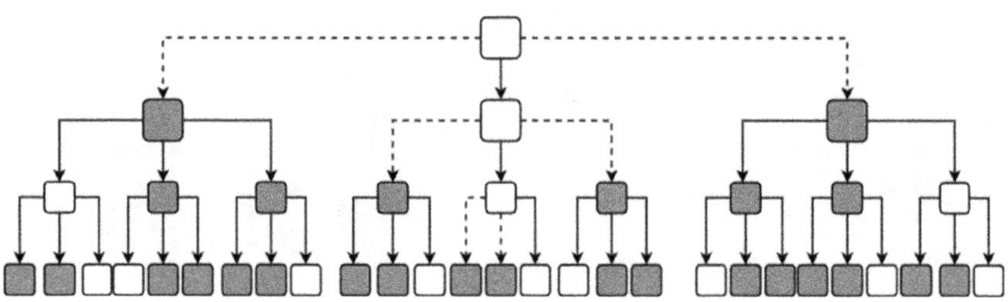

Figure 13: The consequence of immunity in a susceptible population

In figure 13, the thick arrow denotes the disease is transferred, the dotted arrow denotes the disease is not transferred, shaded boxes represent individuals with immunity. To determine the total population, one should rely on the proper government published data and the corresponding updates for the country. If the model expanded for many arbitrary regions, that could also be done for different countries. For example, after getting into the national sample survey in India 2011, if one adopts the birth rate as 20 per year per 1000 individuals (crude birth rate), the birth rate for the proposed model can be computed as follows: BIR=20/365.25/1000 = 5.475702 x 10^{-05}.

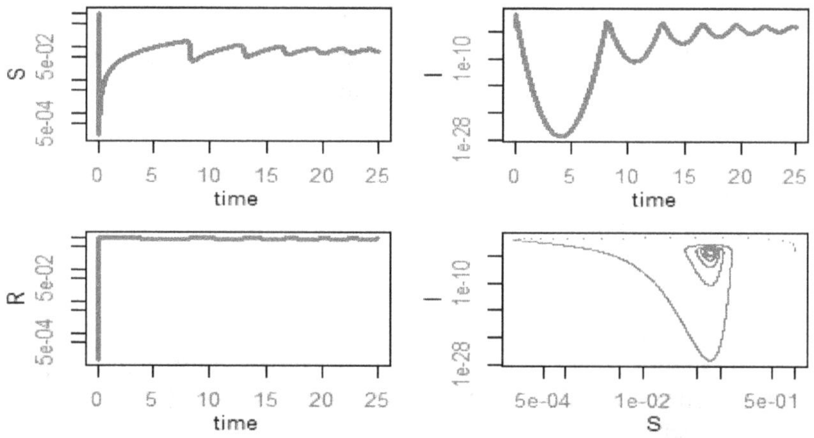

Figure 14: SIR model open population with different IVs

Figure 14 elucidates the SIR technique, including birth, death, and immigration. The disease-induced death rate and natural death rate can be found by relying on either government published data, medical data or previously published articles and literature. It is mandatory to get permission for any scientific publications from the concerned authorities in some data. We have to check the licenses of the secondary data too.

5.3 SIR with seasonal parameter (SP)

Many diseases are seasonal; weather, humidity, and temperature play a crucial role in determining the epidemic. [62] Hence, the earlier said model with incorporated SP is given below.

$$\frac{dS}{dt} = \text{birth}*(1\text{-}S) - \beta * S * I;$$

$$\frac{dI}{dt} = \beta * S * I - \gamma * I - \text{death} * I;$$

$$\frac{dR}{dt} = \gamma * I - \text{death}*R; \qquad \beta(t) = \beta_0\left(1 + \beta_0 Cos 2\pi t\right)$$

Now, one may have the question of getting the appropriate seasonal model for our models. It is suggested to rely on pieces of literature of [62], [63], [64], [65] to explore various seasonal models, and for parameter estimation [66] might be highly useful for the readers. The researcher has to adopt the appropriate model or propose their model for the specific requirement. Figure 14 gives the solution of the DF model with SP, using the program by [36].

```
library(deSolve); seas.sir.model <- function (t, x, params)
 { S <- x[1] ; I <- x[2] ; R <- x[3]
 beta0 <- params["beta0"] ; beta1 <- params["beta1"] ; mu <-
params["mu"]
 gamma <- params["gamma"] ; beta <- be-
ta0*(1+beta1*cos(2*pi*t))
 dSdt <- mu*(1-x[1])-beta*x[1]*x[2] ; dIdt <- beta*x[1]*x[2]-
(mu+gamma)*x[2] ; dRdt <- gamma*x[2]-mu*x[3]
 list(c(dSdt,dIdt,dRdt)) }; params <-
c(mu=1/50,beta0=400,beta1=0.15,gamma=365/13)
xstart <- c(S=0.07,I=0.00039,R=0.92961) ; times <-
seq(from=0,to=30,by=7/365) ; out <- as.data.frame( ode(
func=seas.sir.model, y=xstart, times=times, parms=params ) )
op <- par(fig=c(0,1,0,1),mfrow=c(2,2),
mar=c(3,3,1,1),mgp=c(2,1,0))
plot(S~time,data=out,type='l',log='y',col.axis="red",col="red",lw
d=1,xlab="Time (t)")
plot(I~time,data=out,type='l',log='y',col.axis="red",col="red",lw
d=1,xlab="Time (t)")
plot(R~time,data=out,type='l',log='y',col.axis="red",col="red",lw
d=1,xlab="Time (t)")
plot(I~S,data=out,log='xy',pch='.',cex=0.5,col.axis="red",col="re
d",lwd=1)
```

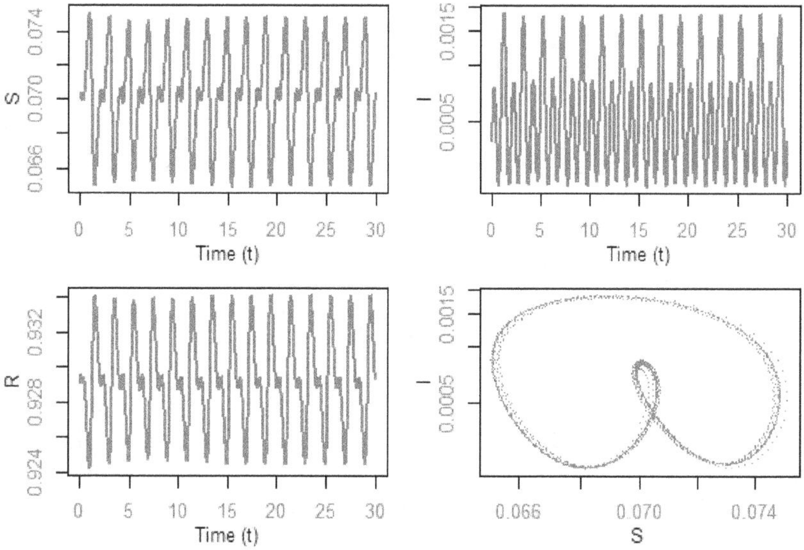

Figure 15: SIR model with SP

If one needs to incorporate a specific parameter, say rainfall, it is better to replace the seasonal with rainfall parameter. For instance, a simple rainfall parameter can be considered as $\log[R_A(t)] = \frac{a * R_A(t)}{b + R_A(t)}$. In 2012, Hii [67] found an autoregressive model for the rainfall and temperature parameter for dengue incidence. The incidence using the ARIMA model is as follows:

$D_{AR} = \phi_0 + \sum_{k=1}^{v} \phi_{t-k} AR_{t-k}$. In 2018, Misra [68] provided a novel approach for simulating rainfall parameters at core monsoon times. To get more insight on the rainfall parameter, one can rely on [68].

People awareness places a vibrant part in controlling the dispersion of diseases. The people with more awareness will have fewer contacts with the disease. Hence, the aware population can be eliminated from the people of S and maintain the total population N, and it will be added with R. Let V_a is the coefficient of awareness. Hence, the model becomes,

$$\frac{\mathrm{dS}}{\mathrm{dt}} = \mathrm{birth} - \beta * S * I - death * S - V_a * S$$

$$\frac{dI}{dt} = \beta * S * I - \gamma * I - \text{death} * I$$

$$\frac{dR}{dt} = \gamma * I + V_a * S - \text{death} * R$$

To get more insight on the calculation of the awareness coefficient, one can rely on [69].

5.4 Inclusion of vaccination

The Risk Ratio (RR) is defined as, "Risk among the unvaccinated group - risk among vaccinated group divided by Risk among the unvaccinated group". The measure in the numerator is called "Risk difference " (RD) or "Excess Risk" (ER) [70], [71]. It is now planned to calculate the Vaccine Effectiveness (VE) with a counter-example. "Assume that in a population of 159, the chickenpox cases are 21, out of which 18 are vaccinated. Out of the remaining non-chickenpox cases, 134 are vaccinated. In an outbreak of chickenpox in Oregon in 2002, chickenpox was diagnosed in 18 of 152 vaccinated children compared with 3 of 7 unvaccinated children. The data was extracted from [72]. To find the VE now." RR = Risk of the syndrome in primary group / Risk of the syndrome in comparison cluster.

RR = Risk of vaccinated / Risk of unvaccinated = (18 / 152) / (3 / 7) = 0.118 / 0.429 = 0.28

VE = (42.9 – 11.8) / 42.9 = 31.1 / 42.9 = 72%

So, if people are immunized, then 72 % chance people have not contracted with the disease than the unvaccinated people. The inoculated cluster got 72% fewer chickenpox cases than if they had not been immunised. Consider the following scenario of S, I and R. Now, the transition probability calculation of DF is given in figure 16.

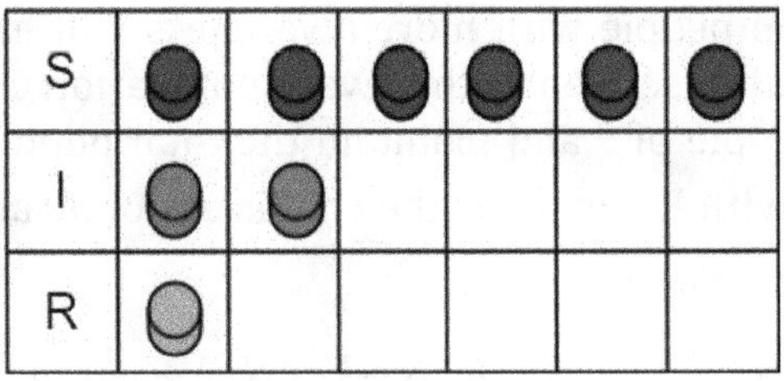

Figure 16: Transition Probability illustration for SIR model

It can be modelled as a transition probability with the vaccine, given in table 2. Here, is incorporated the number of infected neighbours also for assessing possibilities. Assume the vaccination probability as π and average infectious period asτ_i.

Table 2: Transition method and probability for vaccination

Event	Transition from	Transition to	Probability
Infection	s	(s-1,i+1,r)	$1-(1-\beta)^{k_{\inf}}$
Recovery	i	(s,i-1,r+1)	$1-\tau_i$
Vaccine	r	(s-1,I,r+1)	π

The solution of the model after vaccine is given in figure 17 for the initial values 0.9, 0.1 and 0 for S, I and R respectively with birth parameter = 0.001, death parameter = 0.001, vaccination parameter = 0.1 beta value = 0.1, recovery rate = 0.005.

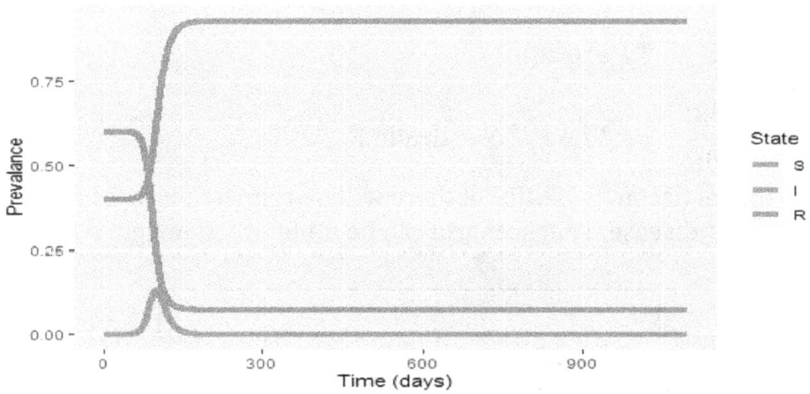

Figure 17: SIR prototype with incorporated vaccine parameter

The S, I, R model can be enhanced with an incorporated state, called "Exposed" (E). Hence, the model becomes the "SEIR" model. SIR model is a deterministic flow model, and the parameter (say ε) denoting the probability of moving from S to E is incorporated. The parameter ε can be determined by calculating the "Speed" or the "Rate of incubation" that leads the exposed state to the infected state. For instance, [73] discovered the incubation period for Covid19; one can get into this for more details. So, it ε tends to infinity, then SEIR becomes SIR. Hence the transformation from the SIR awareness model to SEIR is given in figure 18.

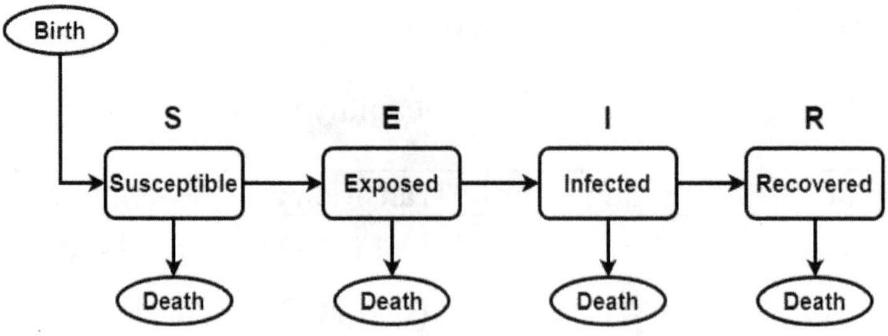

Figure 18: State diagram for SEIR model

$$\frac{dS}{dt} = \text{birth} - \beta * S * I - death * S - V_a * S \quad ;$$

$$\frac{dE}{dt} = \beta * S * I - \varepsilon * E - \text{death} * E$$

$$\frac{dI}{dt} = \varepsilon E - \gamma * I - \text{death} * I \quad ;$$

$$\frac{dR}{dt} = \gamma * I + V_a * S - \text{death} * R$$

The model becomes "SEIRS" if there will be reinfection cases of a particular disease. That scenario can be understood in figure 19.

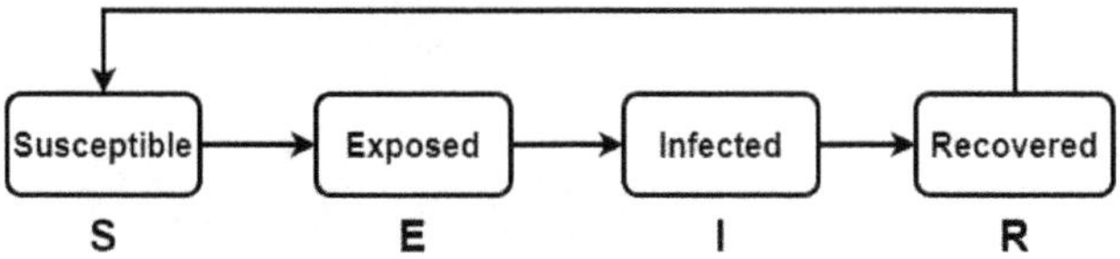

Figure 19: State diagram for SEIRS model

$$\frac{dS}{dt} = \text{birth} - \beta * S * I - death * S - V_a * S + \delta R \quad ;$$

$$\frac{dE}{dt} = \beta * S * I - \varepsilon * E - \text{death} * E$$

$$\frac{dI}{dt} = \varepsilon E - \gamma * I - \text{death} * I \qquad ;$$

$$\frac{dR}{dt} = \gamma * I + V_a * S - \text{death}*R - \delta R$$

5.5 Compartment models

From DF, since the vector, the mosquito is too inved, the model inclusive of the vector can provide more accuracy in the computational results. Figure 20 gives us the compartment for humans with four stages, and figure 21, provides the compartment for mosquitoes with 3 stages, drawn using https://app.diagrams.net/.

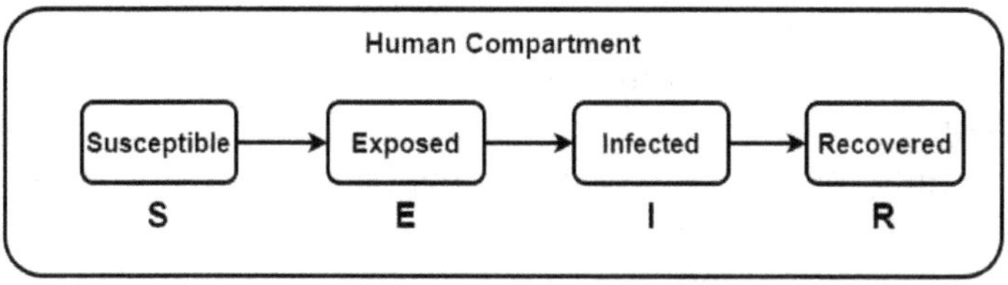

Figure 20: State diagram for SEIR model for Human

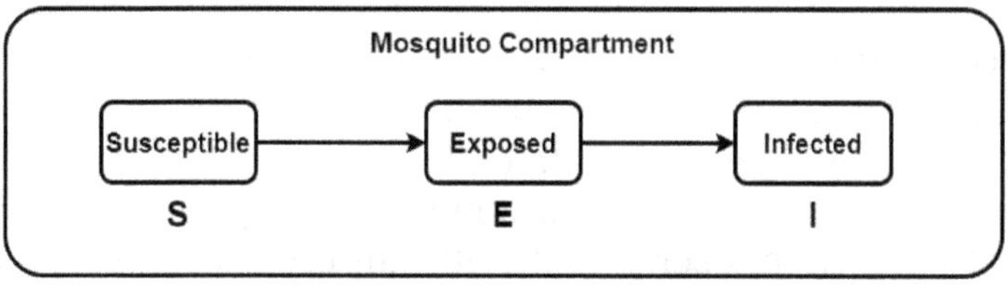

Figure 21: State diagram for SEI model for Mosquito

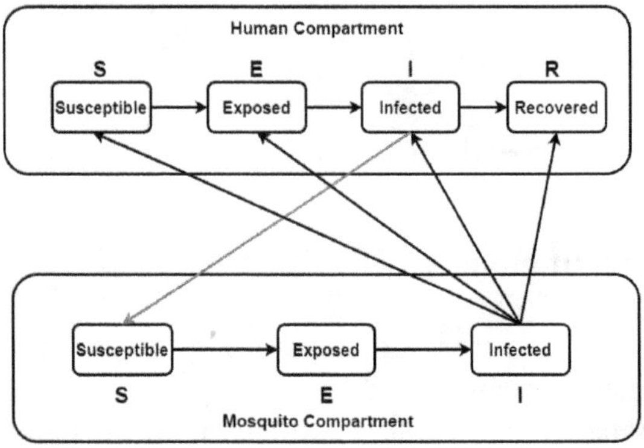

Figure 22: State diagram two-compartment SEIR-SEI model

The connection between the human compartment and the mosquito compartment is given in figure 22. All these figures 20, 21 and 22 are drawn using https://app.diagrams.net/. In the earlier said models, transmit from AAM – H and vice versa is not considered differently. To get a more accurate model, one can consider this as two different quantities and may work on various formulae to determine. Ngwa and Shu [74] proposed their malaria model. [75] proposed Bifurcation Analysis of [74]. [76] proposed spreading models for Malaria with VE incorporated in [75]. The inclusion of immunity for VBD is presented by [77], but the parameters are less than [75]. The recovery rate is split into two parts, with treatment and without treatment. [49] proposed in his DF model that such recovery rate is difficult to analyze since the person with exposing stage will be admitted to the hospital, if there is a mild severity, then it will vanish, and that part will not come to the official records and sometimes it went unnoticed by the people at all. Let $\left[SS\right]_h$ represent the number of susceptible humans, $\left[EX\right]_h$ the number of exposed humans, $\left[IF\right]_h$ the number of infectious humans, $\left[RC\right]_h$ the number of recovered humans t, $\left[SS\right]_m$ the number of susceptible mosquitoes, $\left[EX\right]_m$ the number of exposed mosquitoes, $\left[IF\right]_m$ the number of infectious mosquitoes, $\left[TP\right]_h$ the total number of the human population, and $\left[TP\right]_m$ the total count of AAM at any time *t*.

Then $\left[TP\right]_h = \left[SS\right]_h + \left[EX\right]_h + \left[IF\right]_h + \left[RC\right]_h$ and $\left[TP\right]_m = \left[SS\right]_m + \left[EX\right]_m + \left[IF\right]_m$

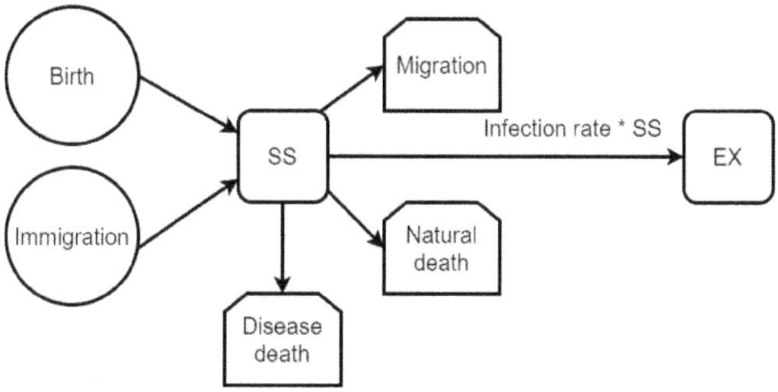

Figure 23: Model creation diagram for SS

The model creation diagram in figure 23, drawn using https://app.diagrams.net/, is proposed for SS. One can follow similar steps for other states and for constructing the rest of the states. Using that methodology, parameters were chosen, the computations and accumulation of parameters to single quantity also have done, yields the following set of equations consists of 7 components. An equation represents every state. The set of all these 7 equations together is the model for the spread of the DF proposed by [78].

$$\frac{d}{dt}\left[SS\right]_h = \wp_h + [BIR]_h [TP]_h + L_h\left[\left[RC\right]_h\right] - \tau_h(t)\left[SS\right]_h - \Omega_h\left[\left[TP\right]_h\right]\left[SS\right]_h$$

In the same manner, one can propose equations for

$$\frac{d}{dt}\left[EX\right]_h, \frac{d}{dt}\left[IF\right]_h, \frac{d}{dt}\left[RC\right]_h, \frac{d}{dt}\left[SS\right]_m, \frac{d}{dt}\left[EX\right]_m \text{and,} \frac{d}{dt}\left[IF\right]_m.$$

Here, for mosquito life span one can consider the literatures [79], [80], [81], and [82]. In 2019, [83] wrote a book on the DF situation in India, from which the readers will get more ideas in creating enhanced models. On the other hand, [84] proposed epidemiological models for the spread of disease. Here, Malaria is taken as a counter-example. The population of an infective human is split into three parts: sensitive untreated, sensitive treated and resistance. The mosquito population is divided into sensitive infected and infected resistance for S and I separately. The mosquito stages can be incorporated as a future development for the model and the primary literature one can rely on [85] in which it is modelled with weather and climate change for the disease transmission. The Spatio-temporal dynamical model can be found with Torres-Sorando L, Rodriguez DJ [86] at which he included visitation duration too with [85].

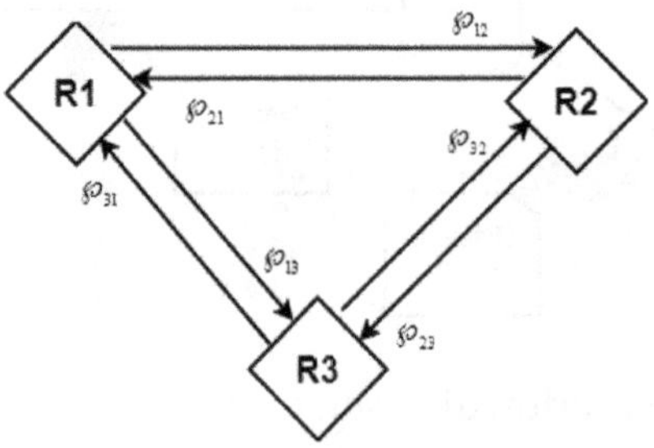

Figure 24: Immigration parameter extension

Immigration can be considered a vital parameter in the estimation in their proposed model. Still, it is usually not suitable to include this parameter in the infected stage, as it is believed that people with infection (symptomatic, hospitalized, under medication) will not travel. Figure 24, drawn using https://app.diagrams.net/, illustrates the immigration parameter addition procedure for the model. It can be generalized, and the models can also be defined for many arbitrary regions. Such models are quite applicable in analyzing the spread demographically. For a heterogeneous environment with spatial separation [87] proposed forecasting Models. [88] has given an algorithm for the stochastic simulation of reactions in chemistry, and one can use this simulation technique for the simulated models of DF. The asymptotic stability can be identified by finding (i) disease-free equilibrium, (ii) creation of Jacobian (iii) finding Eigenvalues and by finding the largest Eigenvalue, which is similar to the threshold for R0. If the Eigenvalues are negative, the disease is stable, and if at least one is positive, it is unstable. Hence one can use [88] to analyze the system modelling and the stochastic behaviour. The Gillespie procedure produces a statistically precise trajectory of a probabilistic ODE too. The method of simulating is (i) to produce uniformly distributed random numbers in the interval 0 to 1, (ii) computation of response interval (iii) accomplish the mth response by changing the pretentious state variables (iv) updating the probabilities $p_s = \sum_{i=n}^{N} p_i$. One can also use other simulation algorithms as equivalent to the Gillespie algorithm. Determination of the final size of the epidemic plays a crucial role to analyze the economic burden, preparatory works by the government, awareness spread, and many more. It is the proportion of population that experience infection by the end of the epidemic. The work of the administrations to reduce the final size, say z. It can be computed using the simple formula $z = 1 - e^{-xR_0}$. Figure 25 illustrates the solution of the final epidemic size with R0=1.83.

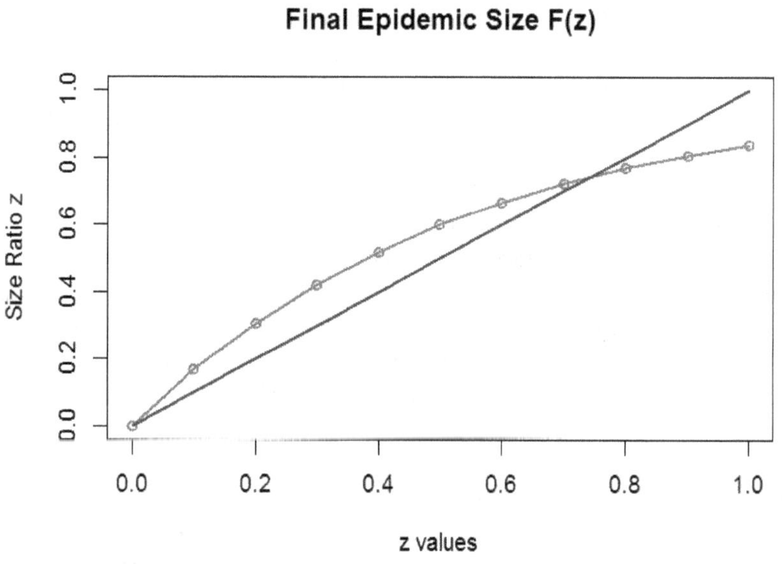

Figure 25: Final size of the Epidemic

The above diagram was drawn with R programming using the following code:
x<-seq(0,1,0.1) ; y<-x ; z<-1-(exp(-x*1.83)); plot(x,z,xlim = c(0,1),ylim = c(0,1),type = "o",col="red",lty=1,lwd=3, xlab = "z values", ylab = "Size Ratio z")
lines(x,y,col="blue",lty=1,lwd=3); title("Final Epidemic Size F(z)")

For an instance, If R0 = 1.83, one can get z = 74% approximately. The graph illustrating the concluding extent of the epidemic is given in figure 25. An extensive discussion on the size of the epidemic was made in the pieces of literature few of the notables are [89], [90], and [91]. The control analysis for DF was proposed by [92]. [93] proposed the solutions of fractional DD non-local systems. A model with fractional order was proposed by [94] in 2021 and [95] in 2022. [95] exponential decay to get the fractional derivatives, the Adam-Bashworth method is adopted for numerical analysis. [96] analysed the model of [94] using the Newton polynomial method. One can develop that model by utilizing various decay models and other numerical analysis models to analyse. [97] developed a smart shield to be used by front line workers, at which it automatically detects the opponent's temperature from a safe distance. On that [97] used nine states with a nonlinear equation for every state viz, S(1), E (2), Asymptotic (2), hospitalized (2), critical (1), R (1) model for the forecasting of infective and to develop the programmatic board. [24] quoted in his material that, to analyse the economic burden of DF, it is analysed the cost sharing of various factors for DF. It reveals the following: Hospitalization 63%, emergency vehicle 17%, non-medical expenses 8% approximately. One may start working on such a global burden of these expenses. Such a task may be categorized, gender, income based and age structured also. For the data collection of various categorized global factors, one can use the "GBD tool http://ghdx.healthdata.org/gbd-results-tool". This tool has many advantages and numerous data and it is left for the researcher, as the data collection method is different for various researchers.

In reality, AAM attacks not just humans but also birds and animals. The DF was also found in mammals and birds. The most severely affected among the birds are "Corvids, crows, and jays"; the rest can carry DenV, however, died in lower quantity. According to [98], horses looked particularly fragile among animals, with a fatality rate of over 40 per cent. An enthusiastic can start their research on the infection model for animals or with specific animal/ bird with age structure, gender structure and many more. The mosquito will lay eggs three days after biting a person and sucking blood, and the cycle will begin again. Female mosquitoes produce many batches of eggs, and the majority of species require a blood meal for each batch laid. Under ideal conditions, an AAM egg can hatch into a larva in less than a day, and an AAM egg can hatch into a larva in less than a day. The larva takes around four days to mature into a pupa, and an adult mosquito emerges after two days. As a result, the mosquito's life cycle may be added for further development of these models. [99] in 2015 proposed the matting pattern of infected mosquitoes, and it can also provide a significant role in the accuracy of the forecasting model.

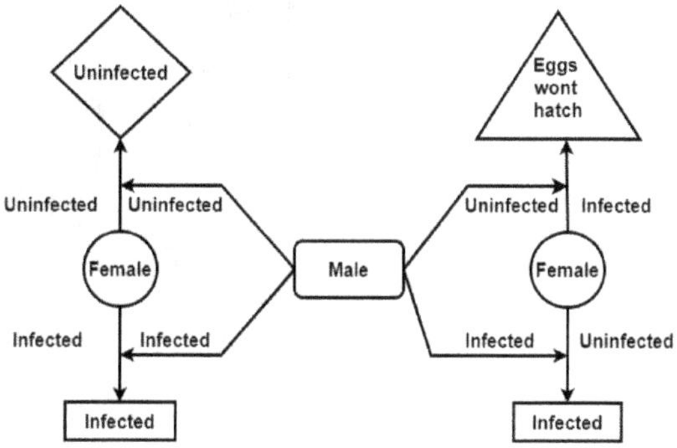

Figure 26: Matting pattern of Mosquitoes

The matting pattern of the mosquitoes can be incorporated for the better accuracy of the model, and the illustration is explained in figure 26, drawn using https://app.diagrams.net/. Many other developments are also possible for the discussed models. For instance, model with reinfection after vaccination of TP, highly sensitive computation for Reductive number, enhanced awareness coefficients, repellents, and other measures, gender-based classification, age-structured and Virus strain variety (four strains).

Figure 27: Word cloud

Zotero software is used to align the references, and the Nvivo tool is used to create the word cloud, treemap of this chapter through automation given in figure 27 and figure 28, respectively.

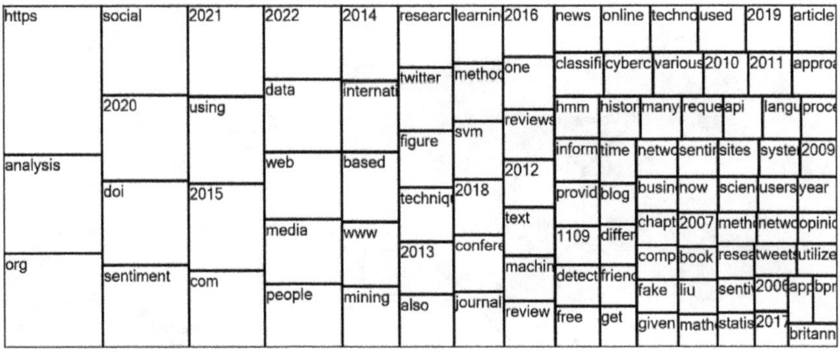

Figure 28: Treemap

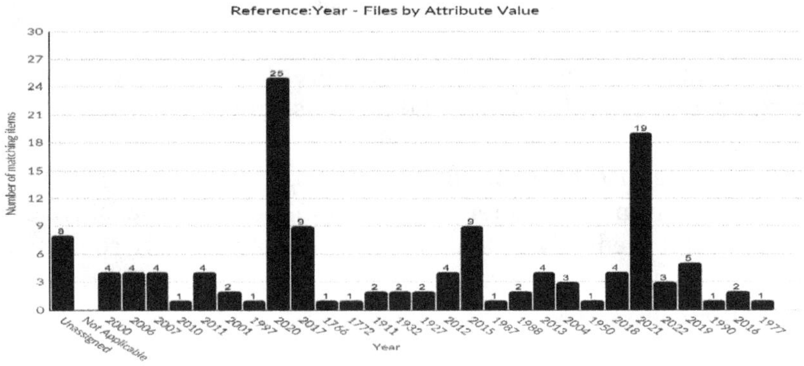

Figure 29: Reference year vs number of articles

The references of this chapter were separately downloaded from Zotero and then uploaded to the standalone nVIVO. The files were categorized and attributed to year-wise classification using the exploration techniques. The automation tool explores the following year wise results of this chapter in figure 29. Researchers, reviewers, or editors can use this tool for their papers/thesis/chapters/books to get into the well-tuned glossary and analyze the recent research vs classical citation ratios. Also, to create the index and the keywords for this chapter, this word cloud method uses nVIVO.

6. Summary

VBD is considered in this chapter to explain the sequential development of the model creation. Here Dengue is taken as a counterexample, the specious AAM acting as a carrier of four strains of the Virus. One may question why we should have this many models instead of keeping one robust model. In this connection, one should believe the statement, "All models are wrong, but some are useful". Alternatively, to develop concrete, more accurate models, preliminary models are necessary. The sequential development of the model from scratch is given and explained lucidly. Many pieces of literature were quoted in between, along with the future results. This chapter possesses the sources and methods of drawing diagrams, the advanced parameter computations are given, and it is explained the necessity of the programming techniques. The modelling flow starts from proposing simple equations through the collected data, creating the ordinary differential equations, proposing the SIR model and the possible parameter incorporations, SEIR model, SEIRS model, SEIR-SEI model, and many future developments. This chapter aims to let the reader get more insight into proposing the models for infectious disease using ordinary differential equations along with probability and new generation parameters. To facilitate the reader, the respective programs too offered within the text to provide the graph. This chapter uses R programming, which the readers can try with any familiar program for their outputs.

References

[1] G. S. Dante and B. William, "disease | Definition, Types, & Control | Britannica," *disease*, Mar. 06, 2020. https://www.britannica.com/science/disease (accessed Jan. 14, 2022).

[2] "Types of Diseases - Study Page." https://www.studypage.in/biology/types-of-diseases (accessed Jan. 14, 2022).

[3] ARTIST and Double Brain, "Disease transmission types-12 vector image on VectorStock," *VectorStock*. https://www.vectorstock.com/royalty-free-vector/disease-transmission-types-12-vector-32352881 (accessed Jan. 14, 2022).

[4] "Types of Diseases Notes | Study Science Class 9 - Class 9," *EDUREV.IN*, Aug. 31, 2017. https://edurev.in/studytube/Types-of-

Diseases/be9703c5-8912-4ff8-8bdf-1b1d27210a6a_t (accessed Jan. 14, 2022).
[5] Kunj, "Viral | Bacterial Human Disease and health for exams," *Sarkari Exam Easy.* https://www.sarkariexameasy.com/2020/05/human-disease-and-health.html (accessed Jan. 14, 2022).
[6] Shaalaa.com, "Non-verbal to Verbal: Transfer the information into a paragraph:Observe the tree diagram of types of disease and write a paragraph on it. Suggest a suitable title. - English | Shaalaa.com." https://www.shaalaa.com/question-bank-solutions/non-verbal-to-verbal-transfer-the-information-into-a-paragraph-observe-the-tree-diagram-of-types-of-disease-and-write-a-paragraph-on-it-suggest-a-suitable-title-writing-skill_203289 (accessed Jan. 14, 2022).
[7] E. WHO, "Dengue vaccine: WHO position paper, September 2018 - Recommendations," *Vaccine,* **37**(35), pp. 4848–4849, Aug. 2019, doi: 10.1016/j.vaccine.2018.09.063.
[8] N. Gupta, S. Srivastava, A. Jain, and U. C. Chaturvedi, "Dengue in India," *Indian J Med Res,* **136**(3), pp. 373–390, Sep. 2012.
[9] E. Massad, F. A. B. Coutinho, L. F. Lopez, and D. R. da Silva, "Modeling the impact of global warming on vector-borne infections," *Physics of Life Reviews,* p. S1571064511000029, Jan. 2011, doi: 10.1016/j.plrev.2011.01.001.
[10] D. Bernoulli, "Essai d'une nouvelle analyse de la mortalité causée par la petite vérole," *Mém. Math Phys Acad Roy Sci Paris.,* **1**(1), pp. 1–45, 1766.
[11] J. Lambert, "Die Toedlichkeit der Kinderblattern. Beytrage zum Gebrauche der Mathematik und deren Anwendung," *Buchhandlung der Realschule,* **3**(3), p. 568, 1772.
[12] R. Ross, *The prevention of Malaria,* 1st ed., 1 s. London : Sterling, VA: John Murray, 1911.
[13] M. McKendrick and W. O. Kermack, "A contribution to the mathematical theory of epidemics," *Proc. R. Soc. Lond. A,* **115**(772), pp. 700–721, Aug. 1927, doi: 10.1098/rspa.1927.0118.
[14] M. McKendrick and W. O. Kermack, "Contributions to the mathematical theory of epidemics. II. —The problem of endemicity," *Proc. R. Soc. Lond. A,* **138**(834), pp. 55–83, Oct. 1932, doi: 10.1098/rspa.1932.0171.
[15] D. L. Smith, K. E. Battle, S. I. Hay, C. M. Barker, T. W. Scott, and F. E. McKenzie, "Ross, Macdonald, and a Theory for the Dynamics

and Control of Mosquito-Transmitted Pathogens," *PLoS Pathog*, **8**(4), p. e1002588, Apr. 2012, doi: 10.1371/journal.ppat.1002588.
[16] S. Dheva Rajan and R. Kalpana, "A study on the chemical applications of Mathematical Graph theory," in *Recent advanced in applied Sciences*, Chennai, Tamilnadu, India, Oct. 2012.
[17] S. Dheva Rajan and R. Kalpana, "A study on the graph theoretical approaches on molecular biology," in *Emerging trends in science and humanities-NCETSH*, Sep. 2012.
[18] S. Dheva Rajan and R. Kalpana, "A study on the graph theoretical properties of chemical molecules," in *Emerging trends in science and humanities-NCETSH*, Sep. 2012.
[19] S. Dheva Rajan, A. Iyem Perumal, and S. Rajagopalan, "A study on Nonparametric statistical linkage for dichotomous traits," in *Challenges in business practices*, Chennai, Tamilnadu, India, Feb. 2012.
[20] S. Dheva Rajan, A. Iyem Perumal, and S. Rajagopalan, "A model for common cold in households," presented at the International Conference on Emerging Trends and Challenges in Science and Technology (ICETCST-2013), Chennai, Tamilnadu, India, Jan. 2013.
[21] S. Dheva Rajan and R. Kalpana, "A Mathematical Model for Natural cooling of a cup of Tea," *Dr.M.G.R PLANETM Online Journal of Mathematical Sciences*, **1**(1), pp. 32–36, 2013.
[22] Y. A. Halasa, B. Zambrano, D. S. Shepard, G. H. Dayan, and L. Coudeville, "Economic Impact of Dengue Illness in the Americas," *The American Journal of Tropical Medicine and Hygiene*, **84**(2), pp. 200–207, Feb. 2011, doi: 10.4269/ajtmh.2011.10-0503.
[23] S. Soh, S. H. Ho, A. Seah, J. Ong, and B. S. Dickens, "Economic impact of dengue in Singapore from 2010 to 2020 and the cost-effectiveness of Wolbachia interventions," *PLOS Glob Public Health*, **1**(10), p. e0000024, Oct. 2021, doi: 10.1371/journal.pgph.0000024.
[24] D. Hariharan, M. K. Das, D. S. Shepard, and N. K. Arora, "Economic burden of dengue illness in India from 2013 to 2016: A systematic analysis," *Int J Infect Dis*, **84**, pp. S68–S73, Jul. 2019, doi: 10.1016/j.ijid.2019.01.010.
[25] T. M. Hung, D. S. Shepard, A. A. Bettis, H. A. Nguyen, and A. McBride, "Productivity costs from a dengue episode in Asia: a systematic literature review," *BMC Infect Dis*, **20**(1), p. 393, Dec. 2020, doi: 10.1186/s12879-020-05109-0.
[26] D. S. Shepard, E. A. Undurraga, Y. A. Halasa, and J. D. Stanaway, "The global economic burden of dengue: a systematic

analysis," *The Lancet Infectious Diseases*, **16**(8), pp. 935–941, Aug. 2016, doi: 10.1016/S1473-3099(16)00146-8.
[27] Z. Zeng, J. Zhan, L. Chen, H. Chen, and S. Cheng, "Global, regional, and national dengue burden from 1990 to 2017: A systematic analysis based on the global burden of disease study 2017," *EClinicalMedicine*, **32**, p. 100712, Feb. 2021, doi: 10.1016/j.eclinm.2020.100712.
[28] N. B. Tjaden, S. M. Thomas, D. Fischer, and C. Beierkuhnlein, "Extrinsic Incubation Period of Dengue: Knowledge, Backlog, and Applications of Temperature Dependence," *PLoS Negl Trop Dis*, **7**(6), p. e2207, Jun. 2013, doi: 10.1371/journal.pntd.0002207.
[29] J. J. Bolhuis, A. M. Strijkstra, and R. J. Kramers, "Effects of scopolamine on performance of rats in a delayed-response radial maze task," *Physiol Behav*, **43**(4), pp. 403–409, 1988, doi: 10.1016/0031-9384(88)90111-4.
[30] R. E. Whitmire, D. S. Burke, A. Nisalak, B. A. Harrison, and D. M. Watts, "Effect of Temperature on the Vector Efficiency of Aedes aegypti for Dengue 2 Virus," *The American Journal of Tropical Medicine and Hygiene*, **36**(1), pp. 143–152, Jan. 1987, doi: 10.4269/ajtmh.1987.36.143.
[31] V. Duong, L. Lambrechts, R. E. Paul, S. Ly, and R. S. Lay, "Asymptomatic humans transmit dengue virus to mosquitoes," *Proc Natl Acad Sci USA*, **112**(47), pp. 14688–14693, Nov. 2015, doi: 10.1073/pnas.1508114112.
[32] N. M. Nguyen *et al.*, "Host and viral features of human dengue cases shape the population of infected and infectious Aedes aegypti mosquitoes," *Proceedings of the National Academy of Sciences*, **110**(22), pp. 9072–9077, May 2013, doi: 10.1073/pnas.1303395110.
[33] L. E. López-Montenegro, A. M. Pulecio-Montoya, and G. A. Marcillo-Hernández, "Dengue Cases in Colombia: Mathematical Forecasts for 2018-2022," *MEDICC Rev*, **21**(2–3), pp. 38–45, Jul. 2019, doi: 10.37757/MR2019.V21.N2-3.8.
[34] Jafaruddin, S. W. Indratno, N. Nuraini, A. K. Supriatna, and E. Soewono, "Estimation of the Basic Reproductive Ratio for Dengue Fever at the Take-Off Period of Dengue Infection," *Computational and Mathematical Methods in Medicine*, **2015**, pp. 1–14, 2015, doi: 10.1155/2015/206131.
[35] R. Taghikhani and A. B. Gumel, "Mathematics of dengue transmission dynamics: Roles of vector vertical transmission and

temperature fluctuations," *Infectious Disease Modelling*, **3**, pp. 266–292, 2018, doi: 10.1016/j.idm.2018.09.003.
[36] S Dheva Rajan, "R PROGRAMMING FOR BEGINNERS - MATHEMATICAL PERSPECTIVE," 2020, doi: 10.13140/RG.2.2.10725.42726.
[37] K. Soetaert, T. Petzoldt, and R. W. Setzer, *deSolve: General solvers for initial value problems of ordinary differential equations (ODE), partial differential equations (PDE), differential algebraic equations (DAE) and delay differential equations (DDE)*. 2016. [Online]. Available: http://cran.r-project.org/web/packages/deSolve/deSolve.pdf
[38] C. Rackauckas and Q. Nie, "DifferentialEquations.jl – A Performant and Feature-Rich Ecosystem for Solving Differential Equations in Julia," *Journal of Open Research Software*, **5**, p. 15, May 2017, doi: 10.5334/jors.151.
[39] H. K. Timothy, *odeintr: C++ ODE Solvers Compiled on-Demand.* 2017. [C++11]. Available: https://CRAN.R-project.org/package=odeintr
[40] W. B. Hans, *pracma: Practical Numerical Math Functions.* 2021. [Online]. Available: https://CRAN.R-project.org/package=pracma
[41] R. R. Alfonso, *rODE: Ordinary Differential Equation (ODE) Solvers Written in R Using S4 Classes.* 2017. [R (≥ 3.3.0)]. Available: https://CRAN.R-project.org/package=rODE
[42] *sundialr: An Interface to "SUNDIALS" Ordinary Differential Equation (ODE) Solvers.* 2021. [Rcpp (≥ 0.12.5)]. Available: https://CRAN.R-project.org/package=sundialr
[43] T. B. Kyle, Bill Gillespie, Charles Margossian, Devin Pastoor, and Bill Denney, *mrgsolve: Simulate from ODE-Based Models.* [Online]. Available: https://CRAN.R-project.org/package=mrgsolve
[44] M. Fidler, M. Hallow, J. Wilkins, and W. Wang, *_RxODE: Facilities for Simulating from ODE-Based Models_*. 2022. [Online]. Available: https://CRAN.R-project.org/package=RxODE
[45] S. Greenhalgh and T. Day, "Time-varying and state-dependent recovery rates in epidemiological models," *Infectious Disease Modelling*, **2**(4), pp. 419–430, Nov. 2017, doi: 10.1016/j.idm.2017.09.002.
[46] M. M. Hassan, Md. A. Kalam, S. Shano, Md. R. K. Nayem, and Md. K. Rahman, "Assessment of Epidemiological Determinants of COVID-19 Pandemic Related to Social and Economic Factors Global-

ly," *J. Risk Finance*, **13**(9), p. 194, Sep. 2020, doi: 10.3390/jrfm13090194.
[47] H. R. Bhapkar, P. N. Mahalle, N. Dey, and K. C. Santosh, "Revisited COVID-19 Mortality and Recovery Rates: Are we Missing Recovery Time Period?," *J Med Syst*, **44**(12), p. 202, Dec. 2020, doi: 10.1007/s10916-020-01668-6.
[48] O. Diekmann, J. A. P. Heesterbeek, and J. A. J. Metz, "On the definition and the computation of the basic reproduction ratio R 0 in models for infectious diseases in heterogeneous populations," *J. Math. Biol.*, **28**(4), Jun. 1990, doi: 10.1007/BF00178324.
[49] S. Dheva Rajan, R. Kalpana, A. Iyem Perumal, and S. Rajagopalan, "SPR_SODE Model for dengue fever," *International Journal of Applied Mathematical & Statistical Sciences*, **2**(3), pp. 41–46, Jul. 2013.
[50] J. Baum, G. Pas, and R. Carter, "The R0 journey: from 1950s malaria to COVID-19," *Nature*, **582**(7813), pp. 488–488, Jun. 2020, doi: 10.1038/d41586-020-01882-9.
[51] R. Breban, R. Vardavas, and S. Blower, "Theory versus Data: How to Calculate R0?," *PLoS ONE*, **2**(3), p. e282, Mar. 2007, doi: 10.1371/journal.pone.0000282.
[52] D. Adam, "A guide to R — the pandemic's misunderstood metric," *Nature*, **583**(7816), pp. 346–348, Jul. 2020, doi: 10.1038/d41586-020-02009-w.
[53] P. L. Delamater, E. J. Street, T. F. Leslie, Y. T. Yang, and K. H. Jacobsen, "Complexity of the Basic Reproduction Number," *Emerg. Infect. Dis.*, **25**(1), pp. 1–4, Jan. 2019, doi: 10.3201/eid2501.171901.
[54] K. Janssen and P. Bijma, "The economic value of R0 for selective breeding against microparasitic diseases," *Genet Sel E*, **52**(1), p. 3, Dec. 2020, doi: 10.1186/s12711-020-0526-y.
[55] Q. Li *et al.*, "Early Transmission Dynamics in Wuhan, China, of Novel Coronavirus–Infected Pneumonia," *N Engl J Med*, **382**(13), pp. 1199–1207, Mar. 2020, doi: 10.1056/NEJMoa2001316.
[56] N. M. Nguyen *et al.*, "Host and viral features of human dengue cases shape the population of infected and infectious Aedes aegypti mosquitoes," *Proc. Natl. Acad. Sci*, **110**(22), pp. 9072–9077, May 2013, doi: 10.1073/pnas.1303395110.
[57] F. F. Yap and M. Yong, "Implementation of An Online COVID-19 Epidemic Calculator for Tracking the Spread of the Coronavirus in Singapore and Other Countries," Epidemiology, preprint, Jun. 2020. doi: 10.1101/2020.06.02.20120188.

[58] Álvaro Díez, "Infectious Disease Calculator | Simulate Any Infectious Diseases," Nov. 03, 2020. https://www.omnicalculator.com/health/viral-infection-sir (accessed Jan. 18, 2022).
[59] https://www.aphis.usda.gov/, "VS Outbreak Surveillance Toolbox," *VS Outbreak Surveillance Toolbo*, 2020. https://www.aphis.usda.gov/animal_health/surveillance_toolbox/calculators/prevalence_estimate_calculator.html (accessed Jan. 18, 2022).
[60] CHRISTOPHER CHENEY and HealthLeaders, "New Online Calculator Estimates COVID-19 Mortality Risk," Dec. 18, 2020. https://www.healthleadersmedia.com/clinical-care/new-online-calculator-estimates-covid-19-mortality-risk (accessed Jan. 18, 2022).
[61] F. F. Yap and M. Yong, "Implementation of a real-time, data-driven online Epidemic Calculator for tracking the spread of COVID-19 in Singapore and other countries," *Infect. Dis. Model.*, **6**, pp. 1159–1172, 2021, doi: 10.1016/j.idm.2021.10.002.
[62] N. C. Grassly and C. Fraser, "Seasonal infectious disease epidemiology," *Proc. R. Soc. B.*, **273**(1600), pp. 2541–2550, Oct. 2006, doi: 10.1098/rspb.2006.3604.
[63] D. Osthus, K. S. Hickmann, P. C. Caragea, D. Higdon, and S. Y. Del Valle, "Forecasting seasonal influenza with a state-space SIR model," *Ann. Appl. Stat.*, **11**(1), Mar. 2017, doi: 10.1214/16-AOAS1000.
[64] G.-Q. Sun, Z. Bai, Z.-K. Zhang, T. Zhou, and Z. Jin, "Positive Periodic Solutions of an Epidemic Model with Seasonality," *The Scientific World Journal*, **2013**, pp. 1–10, 2013, doi: 10.1155/2013/470646.
[65] S. Rosa and D. F. M. Torres, "Parameter Estimation, Sensitivity Analysis and Optimal Control of a Periodic Epidemic Model with Application to HRSV in Florida," *Stat., optim. inf. comput.*, **6**(1), pp. 139–149, Feb. 2018, doi: 10.19139/soic.v6i1.472.
[66] L. Wang, Y. Zhou, J. He, B. Zhu, and F. Wang, "An epidemiological forecast model and software assessing interventions on COVID-19 epidemic in China," Infectious Diseases (except HIV/AIDS), preprint, Mar. 2020. doi: 10.1101/2020.02.29.20029421.
[67] Y. L. Hii, H. Zhu, N. Ng, L. C. Ng, and J. Rocklöv, "Forecast of Dengue Incidence Using Temperature and Rainfall," *PLoS Negl Trop Dis*, **6**(11), p. e1908, Nov. 2012, doi: 10.1371/journal.pntd.0001908.

[68] U. Misra, A. Deshamukhya, S. Sharma, and S. Pal, "Simulation of Daily Rainfall from Concurrent Meteorological Parameters over Core Monsoon Region of India: A Novel Approach," *Advances in Meteorology*, **2018**, pp. 1–18, Jul. 2018, doi: 10.1155/2018/3053640.
[69] "Principles of Epidemiology | Lesson 3 - Section 6," Dec. 20, 2021.
https://www.cdc.gov/csels/dsepd/ss1978/lesson3/section6.html (accessed Jan. 17, 2022).
[70] "Principles of Epidemiology | Lesson 3 - Section 6," Dec. 20, 2021.
https://www.cdc.gov/csels/dsepd/ss1978/lesson3/section6.html (accessed Jan. 17, 2022).
[71] R. Doll and A. B. Hill, "Smoking and Carcinoma of the Lung," *BMJ*, **2**, pp. 739–748, Sep. 1950, doi: 10.1136/bmj.2.4682.739.
[72] B. D. Tugwell, L. E. Lee, H. Gillette, E. M. Lorber, K. Hedberg, and P. R. Cieslak, "Chickenpox Outbreak in a Highly Vaccinated School Population," *Pediatrics*, **113**(3), pp. 455–459, Mar. 2004, doi: 10.1542/peds.113.3.455.
[73] J. A. Backer, D. Klinkenberg, and J. Wallinga, "Incubation period of 2019 novel coronavirus (2019-nCoV) infections among travellers from Wuhan, China, 20–28 January 2020," *Eurosurveillance*, **25**(5), Feb. 2020, doi: 10.2807/1560-7917.ES.2020.25.5.2000062.
[74] G. A. Ngwa and W. S. Shu, "A mathematical model for endemic malaria with variable human and mosquito populations," *Mathematical and Computer Modelling*, **32**(7–8), pp. 747–763, Oct. 2000, doi: 10.1016/S0895-7177(00)00169-2.
[75] Nakul Chitnis, J. Cushing, and J. Hyman, "Bifurcation Analysis of a Mathematical Model for Malaria Transmission," *SIAM J Appl Math*, **67**(1), pp. 24–45, 2006.
[76] H. M. Yang, "Malaria transmission model for different levels of acquired immunity and temperature-dependent parameters (vector)," *Rev. Saúde Pública*, **34**(3), pp. 223–231, Jun. 2000, doi: 10.1590/S0034-89102000000300003.
[77] J. A. N. Filipe, E. M. Riley, C. J. Drakeley, C. J. Sutherland, and A. C. Ghani, "Determination of the Processes Driving the Acquisition of Immunity to Malaria Using a Mathematical Transmission Model," *PLoS Comput Biol*, **3**(12), p. e255, Dec. 2007, doi: 10.1371/journal.pcbi.0030255.
[78] Dhevarajan Srinivasavaradhan, "Developing Mathematical model for an Infectious disease," 2020, doi: 10.13140/RG.2.2.32955.23845.

[79] CDC, "Aedes aegypti and Ae. albopictus Mosquito Life Cycles | CDC," *Centers for Disease Control and Prevention*, Mar. 05, 2020. https://www.cdc.gov/mosquitoes/about/life-cycles/aedes.html (accessed Jan. 19, 2022).

[80] D. Goindin, C. Delannay, C. Ramdini, J. Gustave, and F. Fouque, "Parity and Longevity of Aedes aegypti According to Temperatures in Controlled Conditions and Consequences on Dengue Transmission Risks," *PLoS ONE*, **10**(8), p. e0135489, Aug. 2015, doi: 10.1371/journal.pone.0135489.

[81] K. C. Ernst, K. R. Walker, P. Reyes-Castro, T. K. Joy, and A. L. Castro-Luque, "Aedes aegypti (Diptera: Culicidae) Longevity and Differential Emergence of Dengue Fever in Two Cities in Sonora, Mexico," *J Med Entomol*, **54**(1), pp. 204–211, Jan. 2017, doi: 10.1093/jme/tjw141.

[82] A. P. V. Posidonio, L. H. G. Oliveira, H. L. Rique, and F. C. Nunes, "The longevity of Aedes aegypti mosquitoes is determined by carbohydrate intake," *Arq. Bras. Med. Vet. Zootec.*, **73**(1), pp. 162–168, Feb. 2021, doi: 10.1590/1678-4162-12080.

[83] S. Dheva Rajan, *ODE Model and Analysis On Dengue Fever in India*. 2019. Accessed: Jan. 30, 2021. [Online]. Available: https://nbn-resolving.org/urn:nbn:de:101:1-2019071605064592177608

[84] J. Koella and R. Antia, "Epidemiological models for the spread of anti-malarial resistance," *Malaria Journal*, **2**(1), p. 3, Feb. 2003, doi: 10.1186/1475-2875-2-3.

[85] P. E. Parham and E. Michael, "Modeling the Effects of Weather and Climate Change on Malaria Transmission," *Environmental Health Perspectives*, **118**(5), pp. 620–626, May 2010, doi: 10.1289/ehp.0901256.

[86] L. Torres-Sorando and D. J. Rodríguez, "Models of spatio-temporal dynamics in malaria," *Ecological Modelling*, **104**(2–3), pp. 231–240, Dec. 1997, doi: 10.1016/S0304-3800(97)00135-X.

[87] D. Rodríguez and L. Torres-Sorando, "Models of Infectious Diseases in Spatially Heterogeneous Environments," *Bulletin of Mathematical Biology*, **63**(3), pp. 547–571, May 2001, doi: 10.1006/bulm.2001.0231.

[88] D. T. Gillespie, "Exact stochastic simulation of coupled chemical reactions," *J. Phys. Chem.*, **81**(25), pp. 2340–2361, Dec. 1977, doi: 10.1021/j100540a008.

[89] P. Magal, O. Seydi, and G. Webb, "Final Size of an Epidemic for a Two-Group SIR Model," *SIAM J. Appl. Math.*, **76**(5), pp. 2042–2059, Jan. 2016, doi: 10.1137/16M1065392.
[90] A. J. Black and J. V. Ross, "Computation of epidemic final size distributions," *Journal of Theoretical Biology*, **367**, pp. 159–165, Feb. 2015, doi: 10.1016/j.jtbi.2014.11.029.
[91] L. Almeida, P.-A. Bliman, G. Nadin, B. Perthame, and N. Vauchelet, "Final size and convergence rate for an epidemic in heterogeneous populations," *Math. Models Methods Appl. Sci.*, **31**(05), pp. 1021–1051, May 2021, doi: 10.1142/S0218202521500251.
[92] M. A. Khan and Fatmawati, "Dengue infection modeling and its optimal control analysis in East Java, Indonesia," *Heliyon*, **7**(1), p. e06023, Jan. 2021, doi: 10.1016/j.heliyon.2021.e06023.
[93] N. Valliammal, C. Ravichandran, and K. S. Nisar, "Solutions to fractional neutral delay differential non-local systems," *Chaos, Solitons & Fractals*, **138**, p. 109912, Sep. 2020, doi: 10.1016/j.chaos.2020.109912.
[94] E. Bonyah, C. W. Chukwu, M. L. Juga, and Fatmawati, "Modeling fractional order dynamics of Syphilis via Mittag-Leffler law," Epidemiology, preprint, Feb. 2021. doi: 10.1101/2021.02.05.21251119.
[95] E. Bonyah, M. L. Juga, C. W. Chukwu, and Fatmawati, "A fractional order dengue fever model in the context of protected travelers," *Alexandria Engineering Journal*, **61**(1), pp. 927–936, Jan. 2022, doi: 10.1016/j.aej.2021.04.070.
[96] Y. Khan, M. A. Khan, Fatmawati, and N. Faraz, "A fractional Bank competition model in Caputo-Fabrizio derivative through Newton polynomial approach," *Alex. Eng. J.*, **60**(1), pp. 711–718, Feb. 2021, doi: 10.1016/j.aej.2020.10.003.
[97] Times News Service, "Smart shield, software developed in Oman to help in fight against COVID-19 - Times of Oman," Apr. 07, 2021. https://timesofoman.com/article/100049-smart-shield-software-developed-in-oman-to-help-in-fight-against-covid-19 (accessed Jan. 18, 2022).
[98] M. J. Wonham, T. de-Camino-Beck, and M. A. Lewis, "An epidemiological model for West Nile virus: invasion analysis and control applications," *Proc. R. Soc. Lond. B*, **271**(1538), pp. 501–507, Mar. 2004, doi: 10.1098/rspb.2003.2608.
[99] I. Morlais, S. E. Nsango, W. Toussile, L. Abate, and Z. Annan, "Plasmodium falciparum Mating Patterns and Mosquito Infectivity

of Natural Isolates of Gametocytes," *PLoS ONE*, **10**(4), p. e0123777, Apr. 2015, doi: 10.1371/journal.pone.0123777.

Chapter 3

Energy-Efficient Routing Protocol based on K Means Clustering for Increased Node Lifetime

Rohit Srivastava
School of Computer Science, University of Petroleum and Energy Studies, Dehradun

Abstract: The wide use of Wireless Sensor Networks (WSNs) is discouraged by the seriously restricted-energy limitations of the singular sensor hubs. This is the motivation behind why a huge piece of the examination in WSNs centers on the improvement of energy-productive directing conventions. In this work, a new protocol called Energy Saving Ad-Hoc on Demand Distance Vector Routing Protocol (ESAODV) is proposed, which seeks energy preservation through energy-adjusted grouping. ESAODV models swap lower energy node with higher energy node from higher traffic to lower traffic if higher energy node is presents other wise use alternate path. To work on the dependability through excess ways in the organization, it is recommended to have the greatest number of ways between the source and the objective. It is important to have a base number of hubs in each excess way. Network unwavering quality is expanded in networks multipath disjoint hubs, where every hub disjoint way has the greatest number of repetitive ways and the base number of hubs in each excess way. In the multi-way network hub disjoint, the dependability is extremely high. The performance evaluation of ESAODV is carried out through simulation tests, which evince the effectiveness of this protocol in terms of network energy efficiency when compared against other well-known protocols.

1. Introduction

A Network is utilized to associate the gadgets for sending and getting the information. Figure 1 shows the straightforward situation in which there are numerous hubs, which are associated with the remote medium. This remote medium may be created by the centralized device like switch, switch or a PC (Abidoye, 2021).

The Wireless Ethernet Compatibility Alliance, an industry-standard gathering advancing interoperability among 802.11 gadgets, likewise calls the 802.11 standard Wireless Ethernet or Wi-Fi. The 802.11 standard offers two strategies for arranging a remote organization specially appointed and foundation. Prior there was a conversation that Wireless organization refers to an organization, wherein every one of the gadgets conveys without the utilization of wired association. Remote organizations are for the most part executed with some sort of far off data transmission framework that utilizes electromagnetic waves, like radio waves; for the transporter and this execution normally happens at the actual level or "layer" of the organization (Sarita Simaiya, 2017).

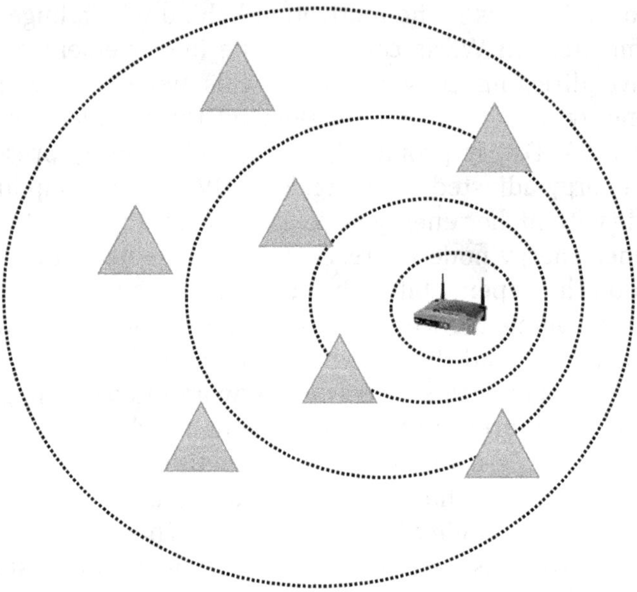

Figure 1: Wireless network

Figure 1 shows the Wireless organization, which have exceptionally restricted transmission capacity and are generally less solid because of ecological impacts in examination of wired organization that give abundant transfer speed and dependable connections in wired organizations. Along these lines, in the remote organization it is lacking that the utilization of conventional organization the board methods, which are utilized in wired organizations.

1.1. Sensor Network

Sensor networks are thick remote organizations of little, minimal expense sensors, which gather and spread ecological information. Already, sensor networks comprised of modest number of sensor hubs that were wired to a focal handling station. Nevertheless, these days, the emphasis is more on remote, conveyed, detecting hubs. When the exact location of a particular phenomenon is unknown, distributed sensing allows for closer placement to the phenomenon than a single sensor would permit. Also, in many cases, multiple sensor nodes are required to overcome environmental obstacles like obstructions, line of sight constraints etc (M. J. Islam, 2007). In most cases, the environment to be monitored does not have an existing infrastructure for either energy or communication. It becomes basic for sensor hubs to make due on little, limited well-springs of energy and impart through a remote correspondence channel.

The arising field of remote sensor networks consolidates detecting, calculation and correspondence into a solitary little unit. With advanced network protocols networking, these gadgets structure an ocean of availability broadens the venture of the internet into the actual world. At the point when water streams occupy each room of a boat lowered lattice network availability will look and work any kind of correspondence information as conceivable hopping starting with one hub then onto the next looking for his fate. While the limit of any single device is minimal (Hong Luo, 2006), the creation of many gadgets offers revolutionary new mechanical conceivable outcomes.

The force of remote sensor networks is the capacity to convey countless little hubs that are gathered and arranged. The situations for the utilization of these gadgets going from constant following, observing natural conditions, to pervasive conditions for in situ wellbeing of designs or hardware checking PC (Srivastava, 2001). Although often called for wireless sensor networks, because they can also control the actuators extend control of cyberspace into the physical world.

The most straightforward utilization of the innovation of remote sensor networks is to screen patterns in information conditions far off low recurrence. For instance, a compound plant can be effectively checked for breaks of many sensors that consequently structure a remote connection point and report the discovery of synthetic holes right away. Unlike traditional cable systems, implementation costs would be minimal. Rather than conveying great many feet of wires directed through the defensive sheath, installers gadget (Nakas, 2020) by just setting quarter size, as outlined in Figure 1 every identification point. The organization could be steadily stretched out by adding different gadgets - not change or complex arrangement. With the gadgets introduced in this proposition, the framework would have the option to screen abnormalities for quite some time on a solitary arrangement of batteries.

As well as essentially diminishing establishment costs, remote sensor networks can progressively adjust to the evolving climate. Coping mechanisms can react to changes in network geography and can make the organization switch between fundamentally various ways. For example, the same board for conducting surveillance on a network leaking chemical plant can be reconfigured in a network to locate the source of a leak, monitoring the spread of toxic gases (Rhim, 2018). The organization could then specialists towards the most secure way for crisis departure. Current remote frameworks just start to expose new freedoms for the reconciliation of low-power correspondence, detecting, energy stockpiling and figuring.
For the most part, when individuals consider remote gadgets they think things, for example, PDAs, PDAs, PCs or 802.11 (Nikolaos A. Pantazis, 2012). These things cost many dollars, target specific applications and depend on the framework stretched out help before sending. Nevertheless, the remote sensor network utilizes little gadgets, inserted minimal expense for a wide scope of uses and is not founded on a previous foundation. The vision is that this inheritance cost less than $ 1 per 2005. Figure 2 shows the situation of the sensor network with the radio scope of every hub.

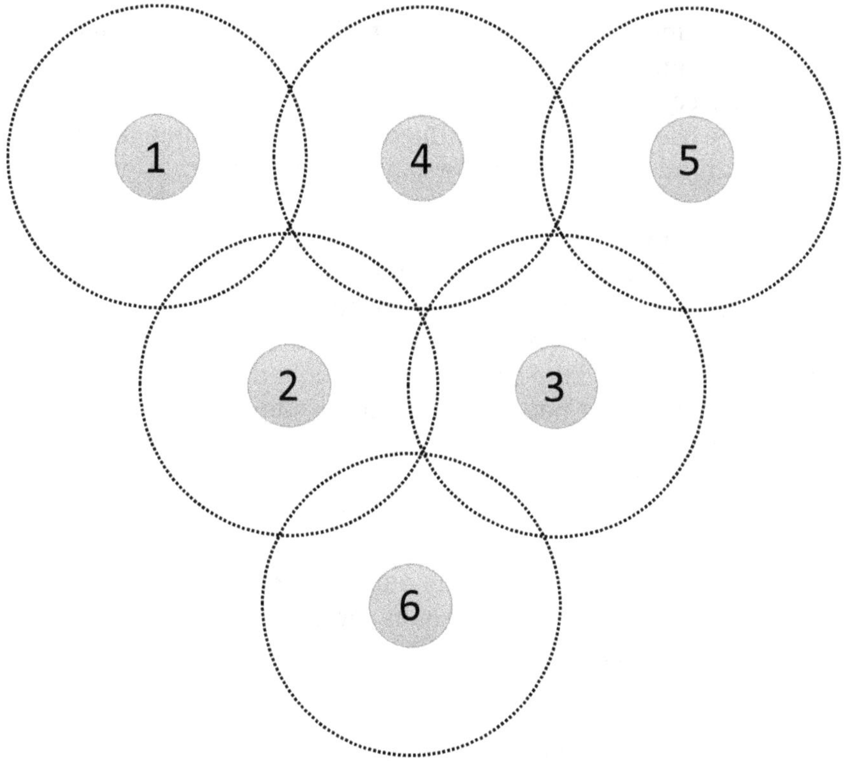

Figure 2: Sensor networks

1.1.1. Applications of Sensor Network

Sensor networks have variety of applications. Models incorporate ecological checking, which includes observing air soil and water, condition-based support, environment observing (deciding the plant and creature species populace and conduct), seismic discovery, military reconnaissance, stock following, savvy spaces and so forth. In fact, due to the pervasive nature of micro-sensors (Obraczka, 2004), sensor networks have the potential to revolutionize the very way we understand and construct the complex physical system.

Because of advances in remote correspondences and hardware as of late, the improvement of organizations of minimal expense, low power, multifunctional sensors has gotten expanding consideration.

The exploration in remote sensor organizations (WSN) is multidisciplinary (Chen, 2013). It contributes to a variety of earrings, such as the scope, equipment, and communication networks in order to achieve the client system issues Ongoing advances in miniature electro-mechanical gadgets for low-power, low-end transmitter, the processor unit low limit and restricted power helped sensor hub plan. WSN is a variation of the specially appointed portable organizations (MANET) comprising of an endless number of little independent gadgets, called remote sensor hubs. An organization of sensors intended to detect events or phenomena, gather and cycle the information and send the data to sink hub identified. The fundamental qualities of sensor networks are densely deployed, the limit with regards to self-association, spread of short-range interchanges, effort agreeable sensor hubs, multi-jump directing, geography changes habitually because of blurring and hub disappointments, restricted energy, power transmission, memory and registering power. WSN (Nikolidakis, 2013) is made out of hundreds or thousands of small sensor hubs. These sensor hubs can speak with one another so the data arrives at receiving hub in a solitary or multi-hop. Sink hub is a typical hub to all objective hubs in the sensor organization.

1.1.2. Challenges in Sensor Network

Dissimilar to customary remote gadgets, remote sensor hubs need not discuss straightforwardly with the control pinnacle or base station closest high power, yet just with their nearby partners. Instead of relying on a pre-deployment each sensor actuator infrastructure, or becomes part of the overall infrastructure. Network protocols peer-to-peer network interconnection provided for transferring data between thousands of tiny embedded devices in multi-hop manner. Flexible mesh architectures considered dynamically adapt to support the introduction of new nodes or extended to cover a wider geographical area. Gadgets in multi-jump way. Moreover, the framework can naturally change in accordance with make up for hub disappointments.

Mesh Network Vision depends on the value of nodes. Dissimilar to cell frameworks that refuse assistance when an excessive number of telephones are dynamic in a little region, interconnect a remote sensor network just develops further as hubs are added. As the density is adequate, a solitary organization hub can be extended to cover the limitless region. Every hub has a correspondence scope of 50 meters and expenses costs less than a sensor network encompassing the Earth Ecuador will cost less than $ 1 million.

1.2. Objective

It has been seen that the greater part of the past approaches for chose alternate path directly way straightforwardly when any hub closure that dropped execution and have relative higher intricacy. As the portable hubs work on the restricted force of battery in this manner it turns out to be exceptionally important to foster methods which can effectively keeping up with lesser intricacy. The target of this work is to foster another methodology which can effectively keep up with the lesser battery power to long endurance of Sensor organization.

1.3 Energy Issues in Wireless Sensor Network

The principle center is around the distributed (and local) designing algorithms, where individual hubs play out their own calculations for figuring answers for a worldwide issue. An appropriated algorithm is one in which the hubs independently execute similar calculation and settle on choices likewise without knowing the overall organization geography. Notwithstanding, in some disseminated calculations, it is passable that the hubs can gain proficiency with some general data (for instance, the quantity of sensors in the Network and/or the greatest level of the basic bend) (Nikolidakis, 2013). A more grounded form of disseminated algorithm is known as the nearby calculation. Informally, a nearby calculation, permits a hub to discuss just with their neighbors, which are in addition to a steady leap away to settle on choices during the execution of the calculation

- Broadcasting issues

  Dispersion is a cycle by which a message created by a hub of the organization, is shipped off any remaining hubs in the organization. Later a straightforward way to deal with flooding is wasteful. This is on the grounds that numerous pointless transmissions (or transmissions just) messages are produced and communicated in the organization, which thus makes the hubs rapidly scatter their valuable energy. Subsequently, we need to plan energy productive calculations that can forestall or if nothing else diminish the amount of redundant transmissions.

- Clustering Bunching

  Bunching is an all-around concentrated on point locally of sensor organizations, where the objective is to separate the whole organization in various gatherings (not really disjoint) and select a hub in the cluster head (CH) of each gathering. Each CH is thought to be dynamic and accomplish practically everything of coordination, eg, recognition, information assortment and information transmission for the benefit of the gathering (S. Prithi, 2020) to the base station, while the other bunch hubs can enter rest mode. In view of collection the issues is to limit the quantity of CHs, in the organization or a CH bunch or possibly straightforwardly associated with CH. That would leave more detectors in low energy reserve. This problem is also known as the minimum set of key problem. In any case, similar to all CHS (even insignificant CHS) is occupied constantly for discovery, handling and transmission of such information, power running rapidly, while different hubs (CHS) are not left with a great deal of energy. This causes a critical power supply hubs and lessens the disequilibrium web of life. One method for tackling the present circumstance is to find a group of disjoint arrangements of CHs and make iterative resources with the goal that the energy utilization is adjusted between the hubs of the organization.

- Observation Target

  Checking (additionally called inclusion) is a significant and broadly concentrated on issue in sensor organizations. In general, the main objective of the research in this area is the design of scheduling algorithms, such as the individual sensors in the network slots to indicate that during the time interval that is active during those days are slots allocated to sleep. Given a WSN that screens specific destinations, it is here and there conceivable to track down a subset of sensors and urge them to do a similar movement checking. So rather than all dynamic hubs for this reason (which is clearly repetitive) we can pick a little subset that can ensure similar management. This observation led researchers to design efficient algorithms so that at any time a small number of nodes are only active for controlling all the objectives in question

- Self-Protect sensor organizations

  The work concentrate on a fascinating issue which manages the arrangement of the sensors with a degree of security by different sensors. Sensors for checking the objective, it is generally expected important to give a degree of security (extra sensors) with the goal that the sensors can make specific moves when attacks are aimed at them. A characteristic thought is to screen sensors by neighbors as neighbors can illuminate the base station when different sensors are at serious risk (or not working because of a glitch). Once in a while whether every one of the sensors in the organization is beneficial to get their work done. From a broken sensor, i.e. a flawed or compromised sensor can't answer to the base station of the condition, targets constrained by the defective sensor become unprotected and the framework has no real way to find out about weakness. For this situation track, down a subset of sensors whose capacity is to control different sensors, so when all sensors (counting themselves) fail or malfunction sensors reported the situation to the base station to the base station. The base station and afterwards make a suitable move, for example, the sending of extra hubs to supplant those that don't work ceaselessly to give insurance to these destinations.

1.4 Literature Survey

According to (Doja, 2012) Remote sensor networks have been created and applied to modern, business, protection and common area applications. Energy is the fundamental obstruction in sensor organizations. Energy management increases the life cycle of the sensor array and improves production efficiency. Ways to deal with multi-jump correspondence and grouping are utilized to save hub energy in sensor organizations. Energy the executives in bunches Self-organized Energy Conscious Clustering (SECC) hub work boundaries (like distant hub, power hub, the hub thickness) and bunch boundaries, (for example, group thickness, sensor hubs per bunch).

Execution examination and recreation results are given with varieties in the number of groups, the energy levels and the separation from the hub.

In the article (Boniewicz, Kozłowska, Zawadzka, Łukasiak, & Zieliński, 2014), the calculations proposed in the strategy for the remote sensor network is thought about. Energy utilization is vital for self-fuelled radio hubs. However, some energy applications balance is more significant. Organizations of remote sensors utilized in enormous regions, for example, farmland or stores comprise many hubs. The customary technique for steering is coordinated to communicate a brief time frame and low energy utilization. The record reveals instances of calculations that might be utilized in the technique for the remote sensor organization. The point of this strategy is the expansion of the organization by means of an informal way choice to limit the scattering of energy in the organization hubs.

According to (Hartwell, 2013), In any case, expanding the accuracy of the sensor builds energy utilization while observing interruption. Models made to recreate an organization, its conventions and information moves, and an entering specialist has demonstrated to be a successful arrangement of instruments to test network conditions and decide the expense of energy.

In the paper (Baghyalakshmi, Ebenezer, & Satyamurty, 2010) Remote sensor organization (WSN) comprises of sensor hubs with minuscule sensors, figuring and remote correspondences abilities. Today the author is tracking down wide application organization and expanding in light of the fact that it permits dependable observing and examination climate. RAP, a correspondence design for sensor networks continuously scale essentially lessening idleness to quit utilizing the maximum velocity-repetitive programming (VMS). APRN, current power steering convention that upholds continuous energy effectiveness constant correspondence by progressively adjusting transmission power and directing choices. The benefits and execution issues of each steering convention are additionally talked about.

As per (Cui, Qu, & Yin, 2013) the energy of the remote sensor organization (WSN) is normally fueled by restricted batteries and convenience. As it's anything but a major test, combined with the developing interest in sensor networks for interactive media applications, the utilization of sustainable power in sensor networks becomes fundamental. In this work, WSN hubs produce environmentally friendly power utilized for directing and identification by the sunlight based charger and a steering convention for energy proficiency energy recuperation. The exploratory outcome shows that the proposed steering: schema-hop routing protocol and power-aware dynamic routing as far as transmission quality, energy utilization and energy use.

According to (Energy Efficient and Fault Tolerant Routing Protocol for Mobile Sensor Network, 2011) the plan of conventions for effective and solid steering for portability situated energy remote sensor organizations (WSN) applications, for example, untamed life checking, front line observation and wellbeing checking is a major test in light of the fact that the organization geography changes oftentimes. The recreation results show conventional FTCP MWSN protocol longer lifetime of the organization, the unwavering quality of the Leach-Mobile and conventions existing Mobile-Leach-improved.

In the work (Kordafshari, Pourkabirian, Faez, & Rahimabadi, 2009), the speed of approach of directing convention is given, since the remaining energy in steering choices. Because of the restricted space of a sensor hub power, energy productive directing is a vital issue in sensor organizations. This technique means to foster a steering convention state nearly without, which can be utilized to course information based on remaining energy nodes. The reproduction results show that the new framework further develops timeframe of realistic usability of around 15% more than the speed of conventional organization convention.

1.5 Proposed Methodology

ESAODV is a proactive hub disjoint multipath directing convention. In ESAODV, WSN is accepted to comprise of a few stages Sti = 1, I 2, ..., l dependent on the number of bounces between the source and objective. The sink is a hub St0 zero. Every hub can speak with the recipient hub is St1. We accept that a hub can speak with hubs on a similar stage Sti and the following stage + 1 yet can't speak with Sti-1 hubs. This tries not to circle ways. At first, all organization hubs have an extremely high worth of the bounce count except for the receiving hub. At first, all hubs have their home over the limit energy level energy. Numerous ways from all hubs to the sink is produced in the development period of the path. In the process of building the packages, Route Connection (RCON) are exchanged between nodes. Every sensor hub communicates the bundle once RCON and keeps up with its own steering table. Assuming that there is no way to the sink hub through the RCON got bundle, then, at that point, the hub processes the parcel RCON. If the way to move from this hub is as of now accessible in the steering table of the hub, then, at that point, the quantity of bounces the parcel is checked. On the off chance that the bounce number of bundles is not exactly the worth of the hub and its leftover energy hop is more prominent than the power limit esteem, then, at that point, it is RCON; in any case the parcel is disposed of. The hub getting the RCON bundle refreshes the RCON parcel. RCON is refreshed with gradual number of jumps by one, refreshes the hub ID prior to adding the hub identifier in the manner. The hub getting the RCON bundle refreshes its steering table as the quantity of jumps and way hub to the beneficiary. Likewise, all hubs in the organization get the RCON parcel and update their steering tables. When they are on the whole various ways are created, the hub disjoint multipath distinguished between the source and objective. When the source node sends the data from the target, extends the FFI trace data between nodes disjoint multipath based and long-tail filled fill residual energy. Assuming away disjoint hub bombs because of the passing of steering hub development or hub, it illuminates the source hub through the Route Error bundle.

The algorithm is implemented by modifying the original AODV source code in NS-2.
Figure 3 shows the initial position of the nodes. This is the scenario of 50 nodes. This is a graphical representation of the simulation.

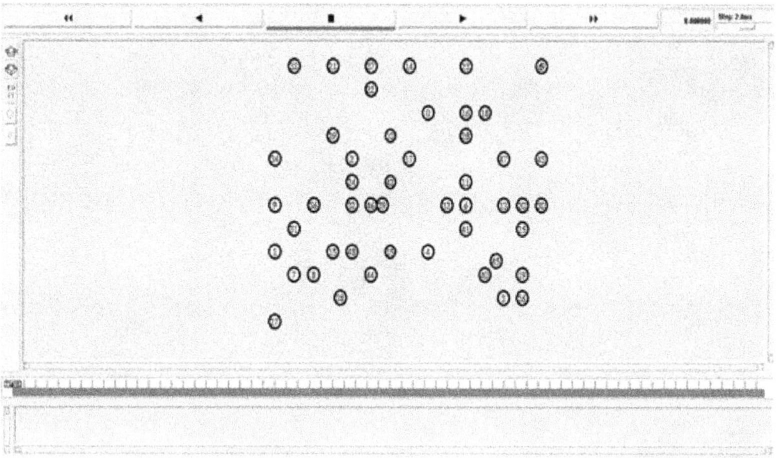

Figure 3: Sensor Network Scenario with 50 nodes

Figure 4 shows the nodes are in the active mode. They are showing their radio range as a circle.

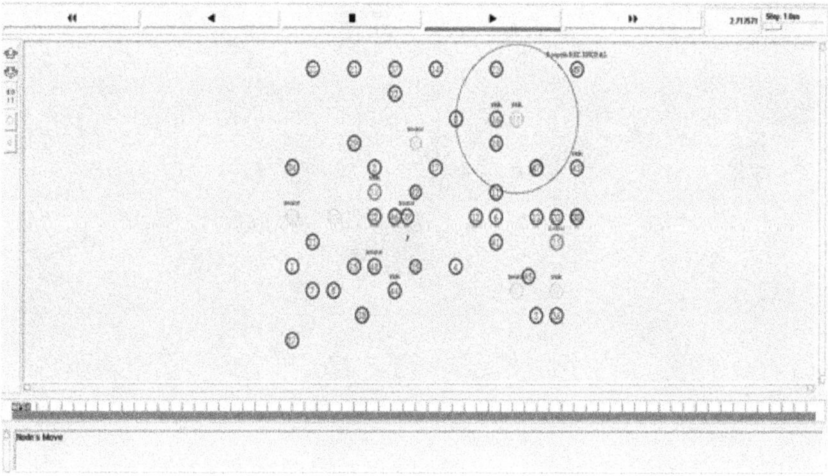

Figure 4: Sensor Network node radio range

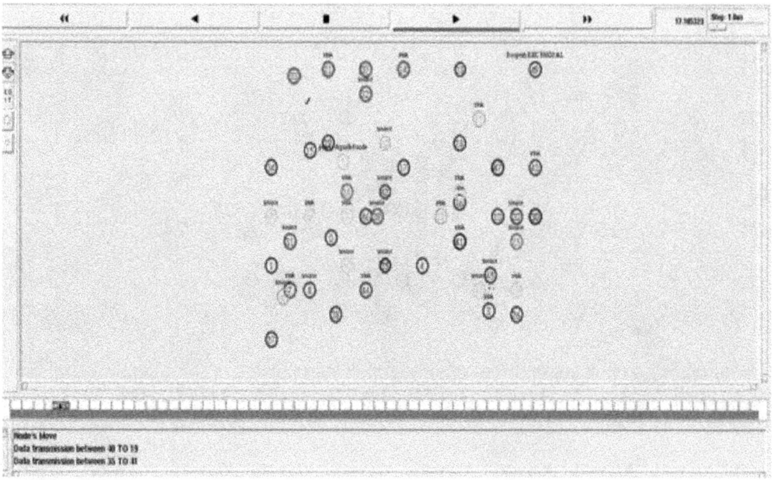

Figure 5: Sensor Network Scenario node with different label

Figure 5 and figure 6 shows sensor network scenario with 50 nodes Here the nodes can freely move in the network. This mobility will change the position of the nodes.

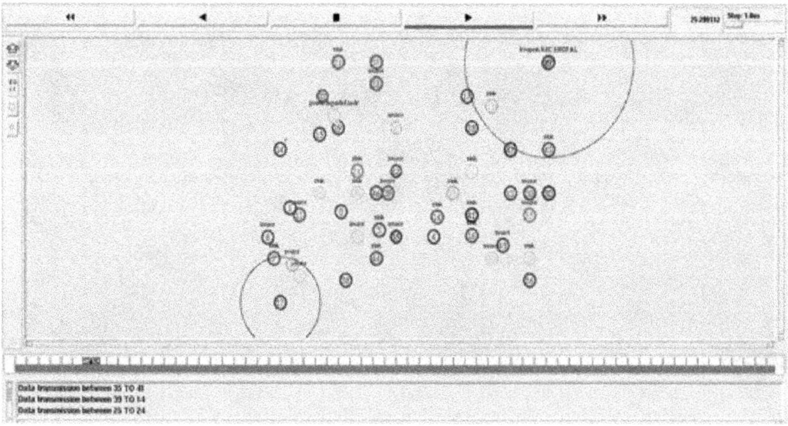

Figure 6: Sensor Network Scenario moving node

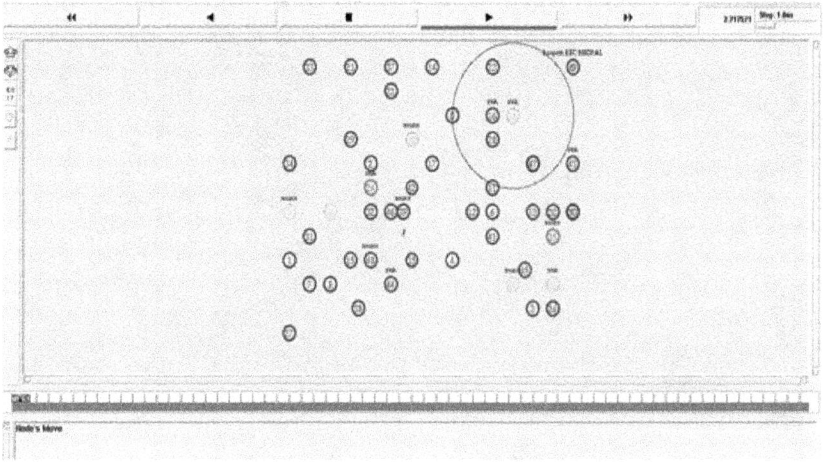

Figure 7: Sensor Network Scenario nodes with data transmission

Figure 7 show data transmission from source node number 21 & 39 towards sink node number 34 & 44 respectively, whereas figure 8 show drop of the data packet at the left bottom.

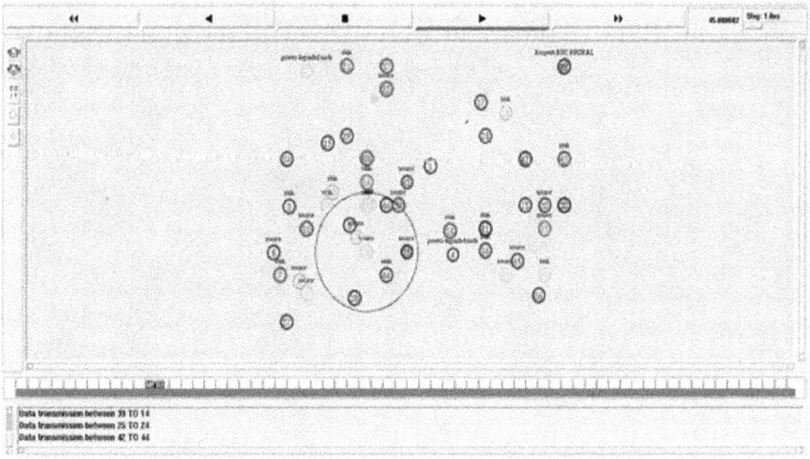

Figure 8: Sensor Network Scenario node with packet drop

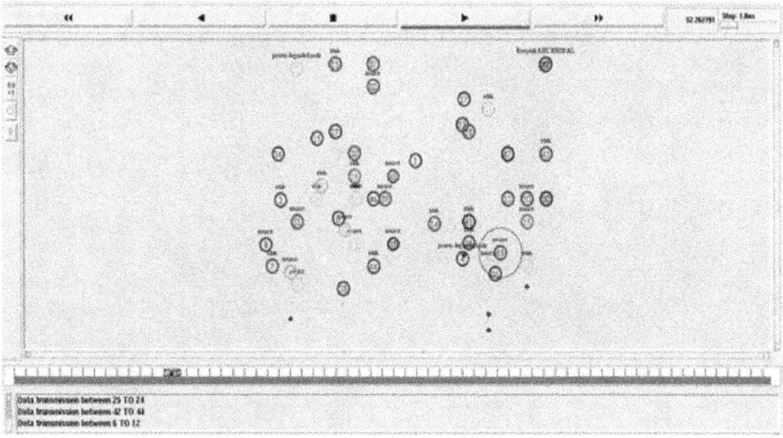

Figure 9: Sensor Network Scenario node with clustering

Figures 9, 10 show node clustering on behalf of energy along with data transmission and packet drop.

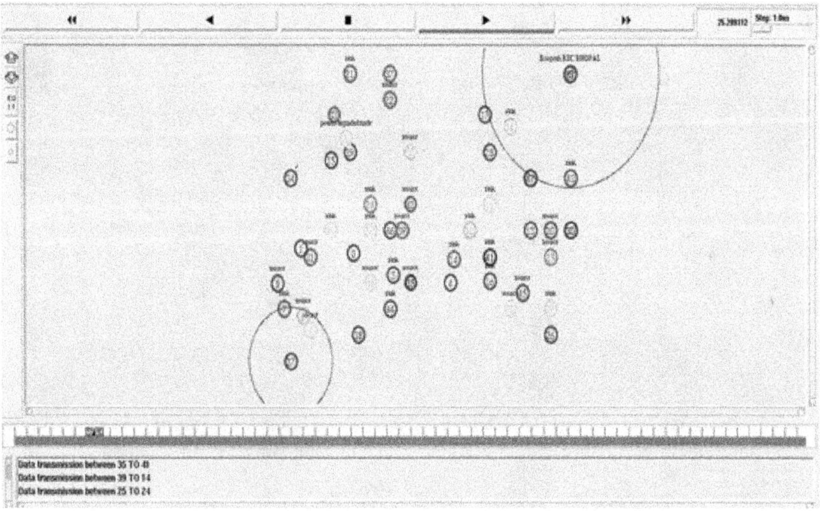

Figure 10: Sensor Network Scenario node with clustering

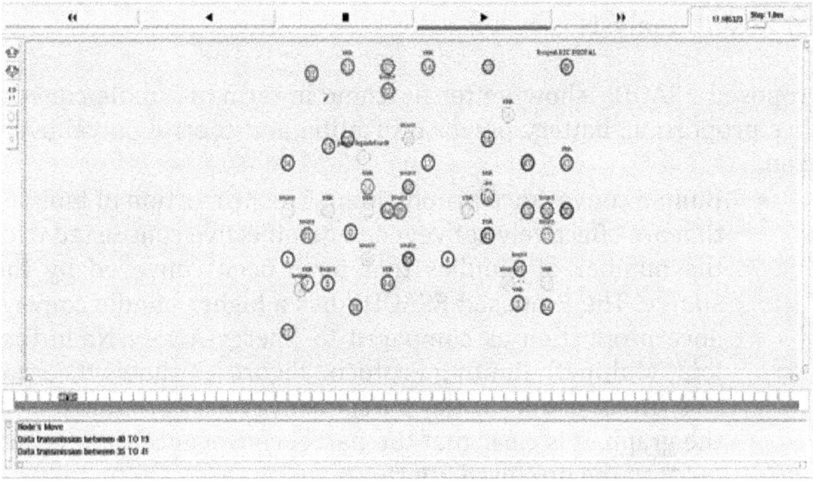

Figure 11: Sensor Network Scenario with discharge node

Figure 11 and 12 show discharge with labels 5 and 23 respectively and show the cluster of lower energy nodes at the right bottom corner.

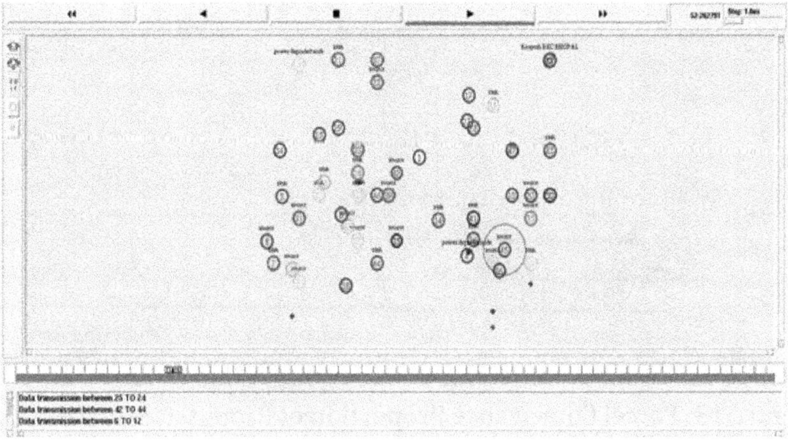

Figure 12: Sensor Network Scenario nodes with clustering

1.6 Result Analysis

Proposed ESAODV show better outcome in term of bundle conveyance proportion, battery power utilization and control parcel overhead.

- Bundle conveyance proportion: - The proportion of bundles that are effectively conveyed to an objective contrasted with the number of bundles that have been conveyed by the source. The Proposed ESAODV has a higher bundle conveyance proportion as compared to Energy-Aware Node Disjoint Multipath Routing Protocol. Figure 13 shows the relative investigation of the parcel conveyance proportion. By the graph, it is clear that the parcel conveyance ratio is better than the proposed approach.

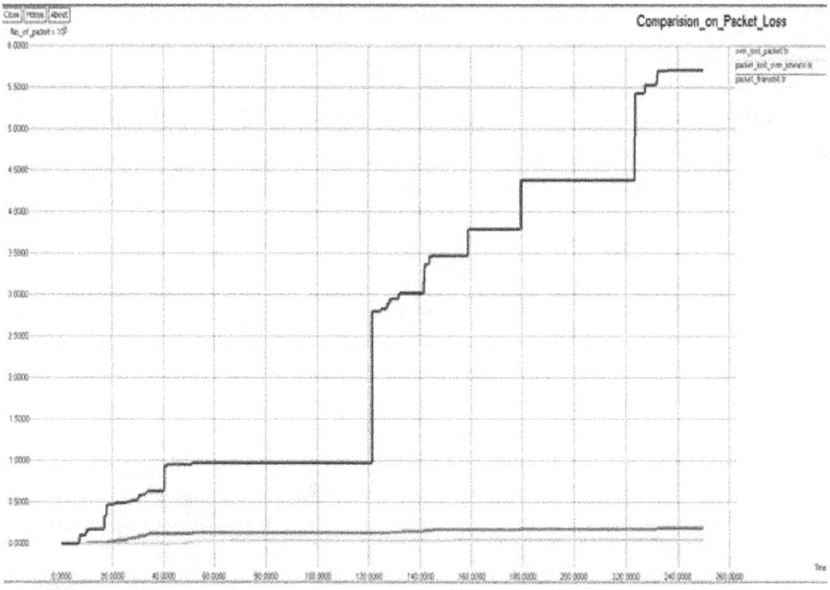

Figure 13: Parcel Conveyance Proportion of Proposed Protocol

- Control bundle overhead: - Bundle parcel overhead is known as a period it takes to send information on a parcel exchanged organization. Every bundle requires additional bytes of organization data that is put away in the parcel header, which, joined with the gathering and dismantling of parcels, lessens the overall transmission speed of the crude information. For any ideal steering, it is required that it has a lower control bundle overhead while existing EENDMRP have required higher control parcel in contrast with the proposed ESAODV.

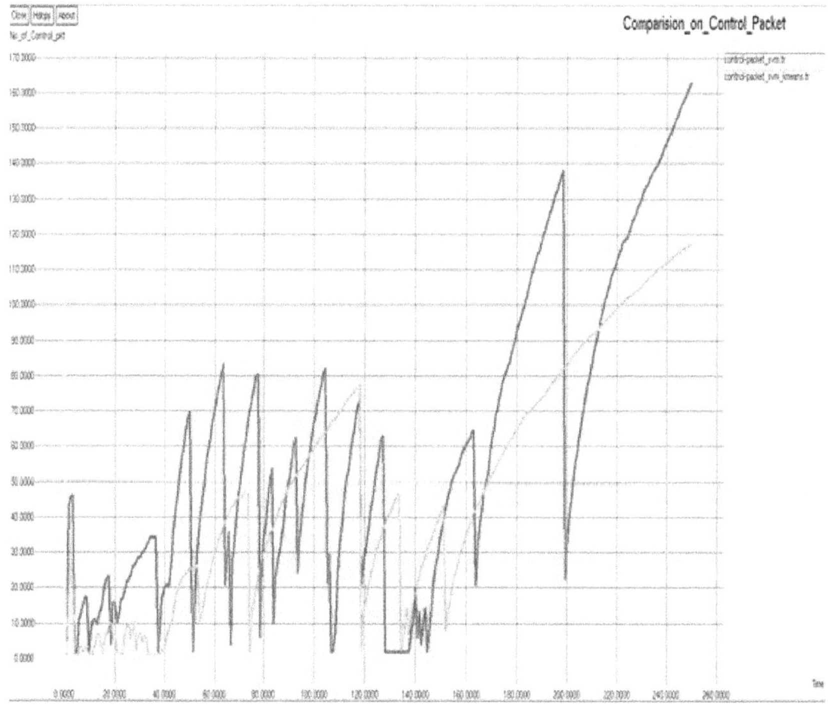

Figure 14: Control Packet Overhead

As far as the graph is concerned it shows the comparative study of the Proposed Protocol and Existing Protocol. The red line shows the previous approach results and the green line shows the proposed approach's result. It is clearly shown that the control packet overhead is less by the proposed methodology.

- Battery Power Consumption:- The energy consumption rate

in a wireless sensor network significantly changes with respect to the protocols used for sensors. Here the sensors are going to use in order to perform the communications. There are various factors like voltage required for the operation; transmission power, received power, etc., that are responsible to calculate the lifetime of a sensor node.

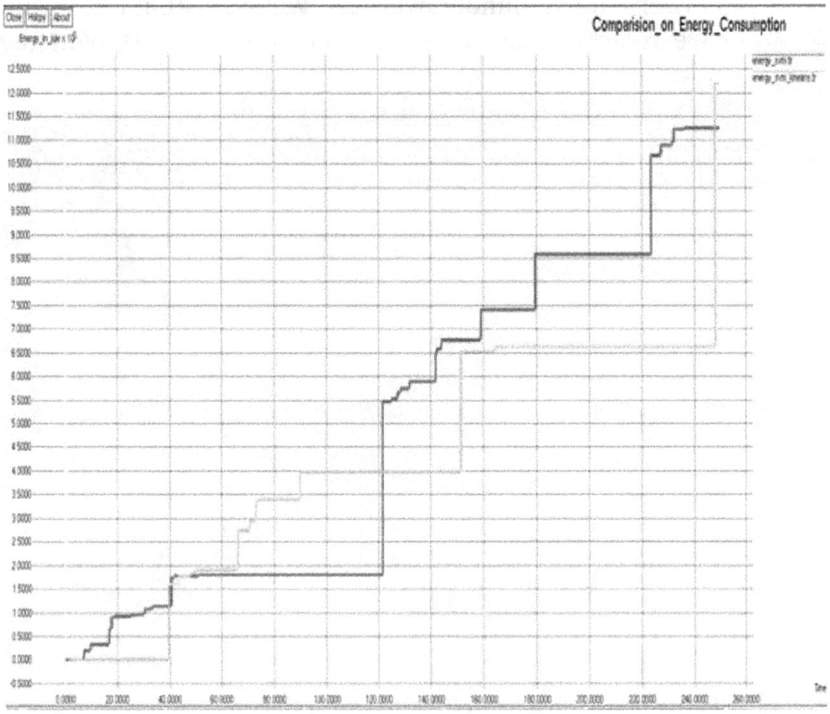

Figure 15: Battery Power Consumption

1.7 Conclusion

The Wireless Sensor Network (WSN) is an arising field for research in the current situation. Flexible mesh architectures are considered progressively adjust to help the new hubs or stretched out to cover a more extensive topographical region. The vast majority of the past approaches for alternate way straightforwardly when any hubs closure that dropped presentation and has relative higher intricacy.

The Mobile hubs additionally have restricted battery power which is difficult for the hub to long time endurance in network.

This work has proposed the ESAODV Protocol for multipath energy productive steering over sensor organization. This technique encapsulate benefit of two distinct predefine strategy to beat their limit. Initial one is substitute way and second one is grouping approach. Proposed protocol attempts to move lower energy hub towards lesser traffic and circulate higher energy hub over weighty traffic segment of organization. In this work reliability through repetitive ways in the organization has been improved, it is recommended to have a most extreme number of ways between the source and the objective. It is important to have a base number of hubs in each repetitive way. Network dependability is expanded in networks multipath disjoint hubs, where every hub has maximum number of paths and the base number of hubs in each repetitive way. In the multi-way network hub disjoint, the unwavering quality is extremely high. The exhibition of proposed procedure is relying on network density and organization traffic. The procedure has been tried through simulations for various appropriations of hubs. Under every one of the assessed situations, the strategy exhibits phenomenal execution in contrast with the existing one. The simulation shows the ESAODV accomplishes significant energy productivity which demonstrates that ESAODV outflanks a few recently proposed conventions like LEACH, PEGASIS and BCDCP.

1.8 Future Scope

In future work, ESAODV can be further enhanced by taking into consideration metrics related to QoS and time constraints. The PDR and throughput can increase and routing load may be decreased with some different approach.

References

1. Abidoye, A. K. (2021). Energy-efficient hierarchical routing in wireless sensor networks based on fog computing. *J Wireless Com Network,* **8**(1), 1687-1499.
2. Baghyalakshmi, D., Ebenezer, J., & Satyamurty, S. (2010). Low latency and energy efficient routing protocols for wire-

less sensor networks. *2010 International Conference on Wireless Communication and Sensor Computing (ICWCSC).* Chennai, India.

3. Boniewicz, M., Kozłowska, A., Zawadzka, A., Łukasiak, Z., & Zieliński, M. (2014). Review of selected algorithms in the method energy evening algorithm in wireless sensor network. *16th International Conference on Transparent Optical Networks (ICTON).* Graz, Austria.
4. Chen, S. W. (2013). LCM: A Link-Aware Clustering Mechanism for Energy-Efficient Routing in Wireless Sensor Networks. *IEEE Sensors Journal,* **13**(2), 728-736.
5. Cui, R., Qu, Z., & Yin, S. (2013). Energy-efficient routing protocol for energy harvesting wireless sensor network. *2013 15th IEEE International Conference on Communication Technology.* Guilin.
6. Doja, M. B. (2012). elf-organized energy conscious clustering protocol for wireless sensor networks. *14th International Conference on Advanced Communication Technology (ICACT).* PyeongChang, Korea.
7. Elena Fasolo, M. R. (2007). In-network Aggregation Techniques for Wireless Sensor Networks: A Survey. *IEEE Communications Surveys,* **1**(4), 1-26.
8. Energy Efficient and Fault Tolerant Routing Protocol for Mobile Sensor Network. (2011). *IEEE International Conference on Communications, ICC 2011.* Kyoto, Japan.
9. Ghaffari, A. (2014). An Energy Efficient Routing Protocol for Wireless Sensor Networks using A-star Algorithm. *Journal of Applied Research and Technology. JART,* **12**(4), 815-822.
10. Hartwell, R. (2013). Wireless Sensor Network Energy Use While Tracking Secure Area Intrusions. *MILCOM 2013 - 2013 IEEE Military Communications Conference.* San Diego, CA, USA.
11. Hong Luo, J. L. (2006). Adaptive Data Fusion for Energy Efficient Routing in Wireless Sensor Networks. *IEEE Transactions on Computers,* **55**(10), 1286-1299.
12. Kordafshari, M. S., Pourkabirian, A., Faez, K., & Rahimabadi, A. M. (2009). Energy-Efficient SPEED Routing Protocol for Wireless Sensor Networks. *2009 Fifth Advanced International Conference on Telecommunications.* Venice/Mestre, Italy.
13. M. J. Islam, M. M. (2007). A-sLEACH: An Advanced Solar Aware Leach Protocol for Energy Efficient Routing in Wire-

less Sensor Networks. *Sixth International Conference on Networking (ICN'07).* Shanghai.
14. Murukesan Loganathan, T. S. (2017). Energy efficient routing protocols for wireless sensor networks: comparison and future directions. *International Conference on Emerging Electronic Solutions for IoT.* Penang, Malaysia.
15. Nakas, C. a. (2020). Energy Efficient Routing in Wireless Sensor Networks: A Comprehensive Survey. *Algorithms,* **13**(3), 72.
16. Nikolaos A. Pantazis, S. A. (2012). Energy-Efficient Routing Protocols in Wireless Sensor Networks : A Survey. *IEEE Communications Surveys,* 1-41.
17. Nikolidakis, S. A. (2013). Energy Efficient Routing in Wireless Sensor Networks Through Balanced Clustering. *Algorithms,* **6**(1), 29-42.
18. Obraczka, I. S. (2004). The impact of timing in data aggregation for sensor networks. *IEEE International Conference in Communications.* Malaysia.
19. Pasquino, N. Y. (2021). Energy-Efficient Routing Protocol Based on Zone for Heterogeneous Wireless Sensor Networks. *Journal of Electrical and Computer Engineering,* **10**(11), 5557756.
20. Rhim, H. T. (2018). A multi-hop graph-based approach for an energy-efficient routing protocol in wireless sensor networks. *Human-centric Computing and Information Sciences,* **8**(1).
21. S. Prithi, S. S. (2020). LD2FA-PSO: A novel Learning Dynamic Deterministic Finite Automata with PSO algorithm for secured energy efficient routing in Wireless Sensor Network. *Ad Hoc Networks,* **97**(1), 102024.
22. Sarita Simaiya, D. S. (2017). Performance And Reliability Improvement Model for WSN. *International Journal of Advanced Research and Publications,* **1**(2), 22-25.
23. Srivastava, C. S. (2001). Energy efficient routing in wireless sensor networks. *MILCOM Proceedings Communications for Network-Centric Operations: Creating the Information Force.* McLean, VA, USA.
24. Subhajit Das, S. B. (2012). Energy Efficient Routing in Wireless Sensor Network. *Procedia Technology,* **6**(1), 731-738.

Chapter 4

Design and Development of Modified Elliptical-Shaped Quad band Microstrip Patch Antenna

Manpreet Kaur
Department of Electronics and Communication Engineering, YDoE, Punjabi University Guru Kashi Campus, Talwandi Sabo, Punjab, India

Abstract: In the current scenario, various domains belonging to the wireless communication sector require compact sized antennas with specific properties. Such antennas are utilized in modern applications due to their compatibility with printed-circuit technology. In this Chapter, the basic objective is to characterize and implement a quad band modified elliptical-shaped microstrip antenna of small size. A commercially available design tool, high-frequency structure simulator (HFSS) has been chosen for the proper designing and efficient simulations. The designed FR4 epoxy based antenna has a size of 37.5 x 37.5 mm^2. In this proposed structure, partial ground plane is used that illustrates its multiband functionality. The feed line bearing an overall size 5.5 x 2.8 mm^2 is connected with the radiating modified elliptical-shaped patch. This Chapter describes the whole structure in detail including all specifications. Afterwards, the antenna is suitably designed with the help of the above-mentioned software and its performance is deeply scrutinized. Several antenna parameters are discussed in this Chapter. From the outcomes, it has been noted that the resultant resonances appear at 4.32, 8.86, 10.10, and 12.15 GHz along with respective S11 value -23.90, -17.32, -23.02, and -13.29 dB. At the respective resultant resonances, the gain values reported are 3.99, 4.16, 7.91, and 6.65 dB. The observed multiple resonances are governed by the specifications of modified elliptical-shaped patch and removed slot, as they can change the current flow. Radiation patterns effectively illustrates the radiation behavior of the realized structure in a better way at the operational frequency bands. Current distribution and radiation efficiency plots are also illustrated about radiation nature. Moreover, performance parameters of the antenna are also evaluated by taking different values of feed width, ground plane length and feeding type. Based on the optimal values, the whole structure shows the best results that claim its suitability for quad band applications.

1. Introduction

Wireless communication was firstly introduced in 19th century and has been continuously in the developing phase. It includes several services, and applications according to the basic requirements of the users [1]. This technology utilizes infra-red, acoustic and any other similar type of waves rather than the wires that can also be used to transmit information [2]. Nowadays, wireless technology basically demands compact and reduced-size antenna devices for the purpose of communication over remotely operated areas [3]. Design specifications of these antennas are very crucial. Such antennas have been employed in wide sectors, such as cellular communication, satellites, defence and health care services, and so on [4]. The most favourable domain of electrodynamics is antenna designing. Antenna constitutes an important part of the communication system. The multiband antennas capable of working at multiple frequencies, while using same resources, are the major need of an hour. Implementation of such high capacity, multi-functional antenna is a challenging activity for today's antenna designers. In recent years, several designs of multiband antennas have been reported and are thoroughly examined [5-9]. In [10], Moore and Gosper curves based antenna was presented for several wireless applications. In [11], several ring-shaped antennas were designed, tested and it was found that these prototypes are useful for wireless applications. Kumar et al. [12] had effectively elaborated the detailed design procedure of a wheel-shaped fractal based device. The selected diameter of the circle was 50 mm. Bangi et al. [13] had revealed the procedure of a compounded fractal antenna that is applicable for multiband applications. In [14], Koch folded-slot antenna was designed and implemented. Nasimuddin et al. [15] had suggested a symmetrically slotted patch antenna. Murugan et al. [16] had illustrated the performance characteristics of a suitably designed circularly polarized antenna. In [17], an antenna having diamond shaped patch with a V-shaped slot etched in the middle was proposed that works at five bands. The structure has a kite shaped slot on it along with C and G alphabet shaped slots in the ground was presented in [18] for quad band antenna. In [19], a fractal shaped Multiple Input Multiple Output (MIMO) antenna was suggested for the most important sub 6-

GHz applications. The proposed structure was flower shaped and supported by defected ground plane. The designed geometry was also experimentally analyzed.

The presented work describes the detailed geometrical specifications and behavior of a modified elliptical-shaped antenna constructed on an adopted square-shaped FR4 substrate material and was backed by a partial ground plane. Focussing on the prime aim, few alterations are carried out onto modified elliptical-shaped antenna design. Microstrip line feed is attached at the middle position with the patch. The introduced antenna is formed, analyzed and optimized using commercial 3D full wave HFSS platform. The Chapter is mainly organized into six sections. It covers the motivation behind this current research. The design procedure along with complete description is given in Section 2 of this Chapter. The results are illustrated in Section 3 with full details. Detailed parametric analysis is presented in Section 4 of the Chapter by considering various cases. Section 5 shows the comparion with other available designs. The whole work is compiled briefly in the last Section of the Chapter.

2. Proposed Antenna

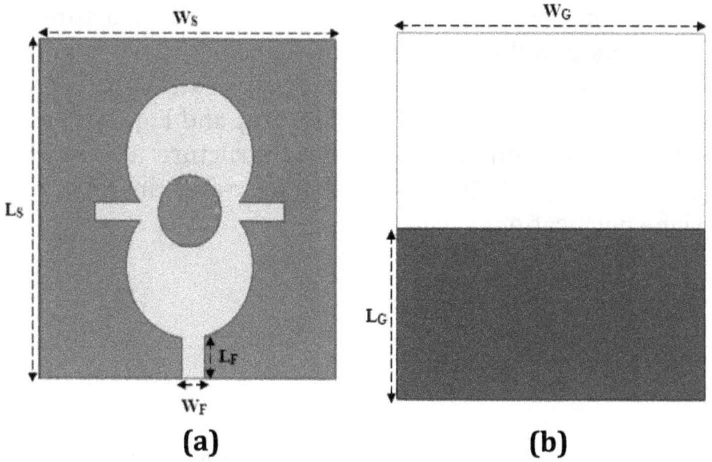

Fig. 1 Proposed antenna: (a) Patch (b) Partial ground plane

Table 1 Design specifications of the suggested antenna

L_S	W_S	L_F	W_F	L_G	W_G
37.5	37.5	5.5	2.8	17.5	37.5
All the parametric dimensions are in mm					

This Section contains deep information about the implementation of modified multiband elliptical-shaped microstrip patch antenna. Concentrating on the specified goal of good radiation behaviour and multiband characteristics, the shape of radiating element selected is modified elliptical. The whole modified elliptical-shaped set up is placed at the centre of the top of square-shaped adopted material of area 37.5 x 37.5 mm^2. The basic characteristics i.e. dielectric constant (ε_r) of the adopted material is 4.4 [28,30]. The designed patch has a unique shape. A circular slot is removed from the patch at the centre. The radius of the removed slit is 4 mm. At the middle portion of the patch, two extra rectangular slits are attached. The microstrip feed has a length 'L_F' =5.5 mm and width 'W_F' = 2.8 mm. The feed is placed at the location where there is suitable impedance matching. The dimensions of the recessed ground plane are 'L_G' = 17.5 mm and 'W_G' = 37.5 mm [31,32]. The chosen partial ground plane enhanced the performance significantly. The geometrical structure of the realized modified multiband elliptical-shaped microstrip patch antenna is illustrated in Fig. 1. Both the sides of the structure including front and back side are delineated in Fig. 1(a) and Fig. 1(b), respectively. Antenna dimensions of the realized structure are computed from equations (1-2)[27,28]. Table 1 demonstrates the realized structure design specifications.

$$f_r = \frac{c}{2L_{eff}\sqrt{\epsilon_{reff}}} \qquad (1)$$

$$W = \frac{v_o}{2f_r}\sqrt{\frac{2}{\epsilon_r+1}} \qquad (2)$$

In the above equations, resonant frequency (f_r), speed of light in air(c), patch length (L)and permittivity (ϵ_{reff})are defined, respectively. Equation (2) gives the value of patch width (W).

3. Results and Discussion

For better understanding, the realization of the designed strucrure is done with HFSS software [26]. Several antenna parameters are examined at various levels and are briefly.

3.1. S parameters and VSWR

The way of association of the input-output terminals with an electrical framework is illustrated effectively by S parameters [17]. In frequency domain, they specify reflection/transmission characteristics. S parameters basically represent the resonance frequency and the bandwidth coverage of the antenna [24]. The resonating frequencies are 4.32, 8.86, 10.10, and 12.15 GHz. The associated S11 value is -23.90, -17.32, -23.02, and -13.29 dB. Fig. 2 delineates the S11 characteristics of proposed antenna. Voltage Standing Wave Ratio (VSWR) illustrates the transfer of RF power from the power source to the antenna through a two-way transmission line [18,33]. At the evaluated resonances, the VSWR values are 1.10, 1.43, 1.62, and 1.79. Fig. 3 represents the VSWR plot of a modified multiband elliptical-shaped patch antenna.

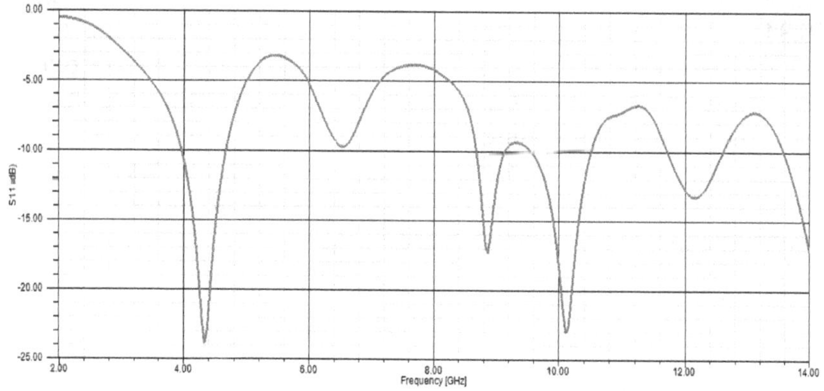

Fig. 2 S11 characteristics of the modified multiband elliptical-shaped patch antenna

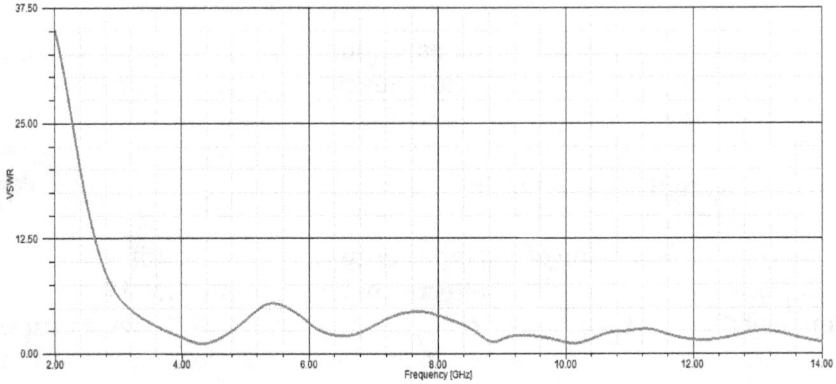

Fig. 3 VSWR plot of the modified multiband elliptical-shaped patch antenna

3.2. Gain and directivity

In an antenna, the metric, gain represents the information about the radiated energy concentration in a particular direction [17,18]. The gain evaluated is ≥ 3 dB at all the associated frequency bands [25,34]. At the resonating frequencies of 4.32, 8.86, 10.10, and 12.15 GHz, the gain values are 3.99, 4.16, 7.91, and 6.65 dB, respectively. The 3-dimensional plots of the modified multiband elliptical-shaped patch antenna at the claimed frequencies are presented in Fig. 4 [35,36]. The realized gain plot in the entire frequency range is shown in Fig. 5. The directivity plot of the modified multiband elliptical-shaped patch antenna is shown in Fig. 6.

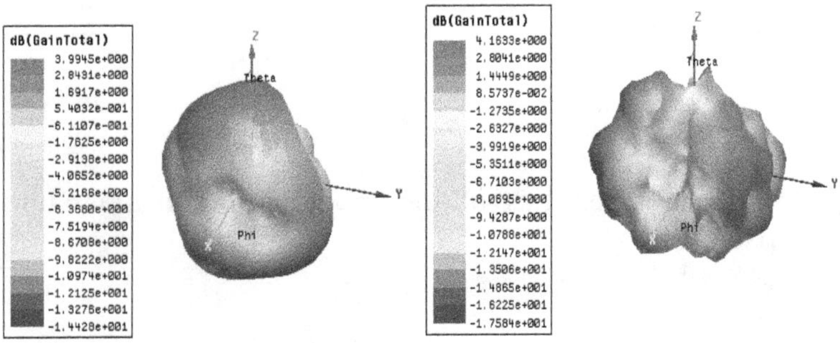

(a) 4.32 GHz (b) 8.86 GHz

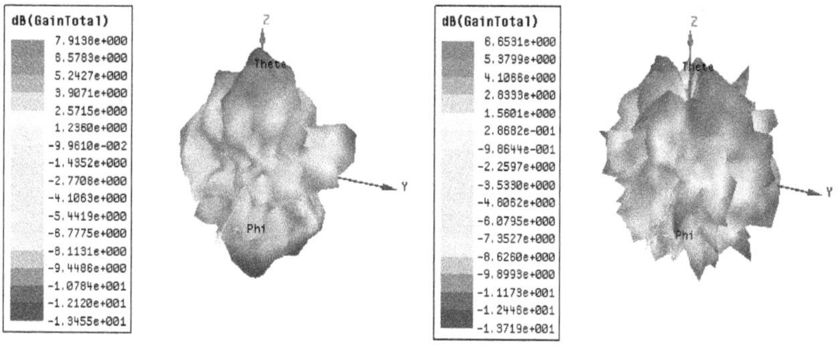

(c) 10.10 GHz (d) 12.15 GHz

Fig. 4 3-dimensional gain plots of the modified multiband elliptical-shaped patch antenna at (a) 4.32 GHz (b) 8.86GHz (c) 10.10 GHz (d) 12.15 GHz

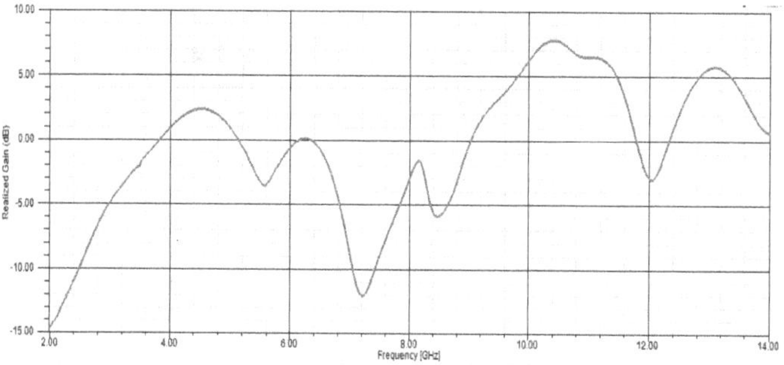

Fig. 5 Realized gain plot of the modified multiband elliptical-shaped patch antenna

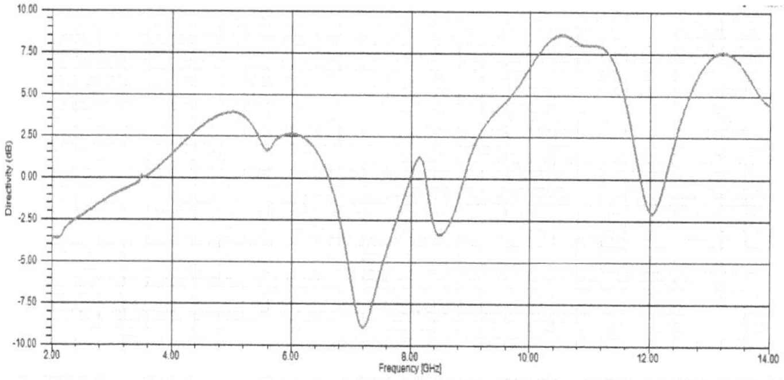

Fig. 6 Directivity plot of the modified multiband elliptical-shaped patch antenna

3.3. Radiation pattern, radiation efficiency and group delay

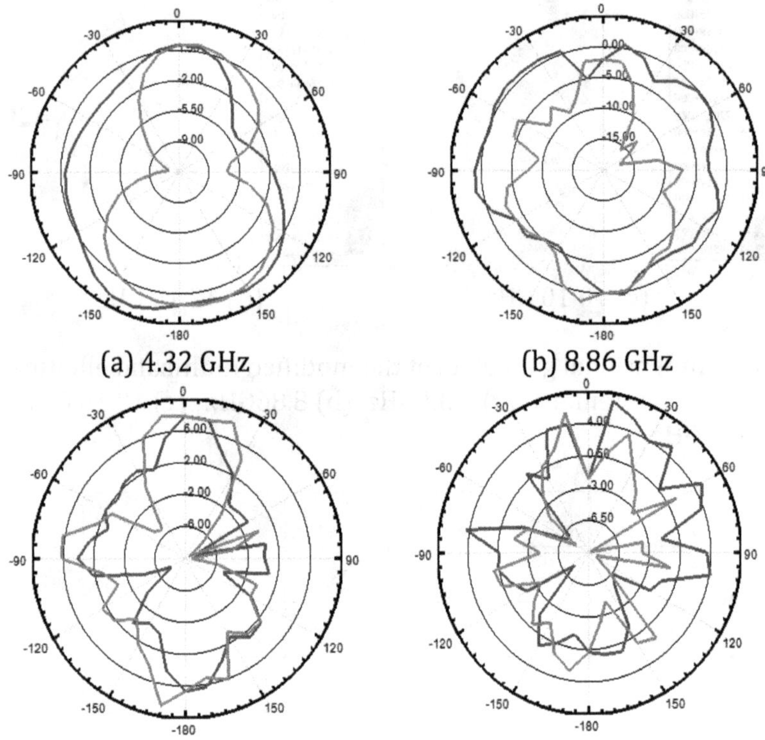

(a) 4.32 GHz (b) 8.86 GHz

(c) 10.10 GHz (d) 12.15 GHz

Fig. 7 Radiation patterns of the modified multiband elliptical-shaped patch antenna at (a) 4.32 GHz (b) 8.86 GHz (c) 10.10 GHz (d) 12.15 GHz

Radiation pattern claims the pictorial representation of the radiated energy distribution in the space with reference to the directional coordinates [17,19]. Fig. 7 illustrates the radiation patterns of the modified multiband elliptical-shaped patch antenna at the investigated resonances. At 4.32 GHz, the pattern observed is bidirectional. At 8.86 GHz, arbitrary omnidirectional behavior is examined. At the third resonance, there is almost bidirectional pattern. At 12.15 GHz, the behavior is arbitrary omnidirectional. Fig. 8 represents the plot of radiation efficiency. At the center frequencies, the radiation efficiency at the respective value is 90%, 68%, 93% and 84%. The value of group delay at the desired resonances is examined from Fig. 9. Table 2 mentions the realized antenna simulated parameters.

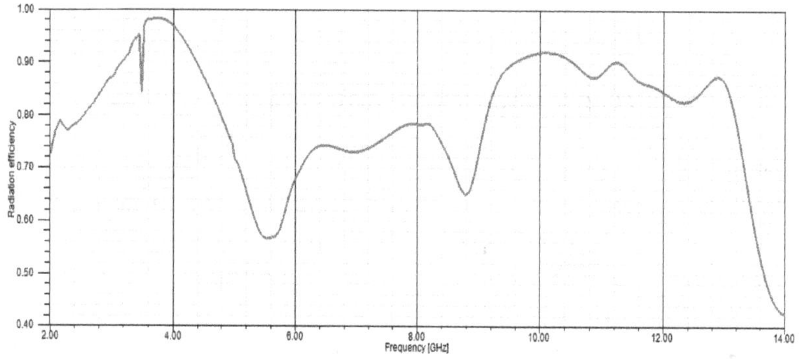

Fig. 8 Radiation efficiency plot of the modified multiband elliptical-shaped patch antenna

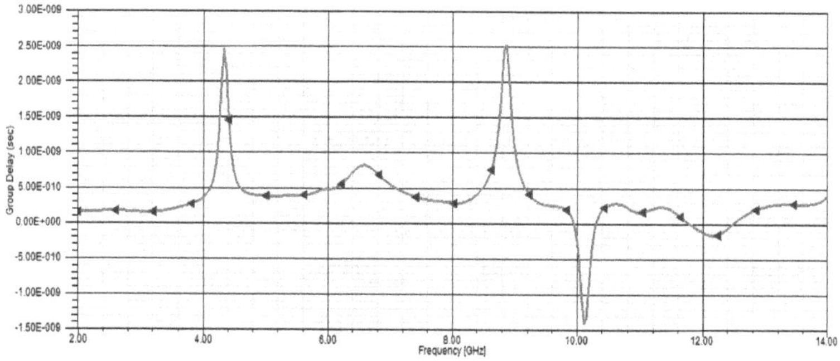

Fig. 9 Group delay plot of the modified multiband elliptical-shaped patch antenna

Table 2 Realized antenna simulated parameters

S. No.	'F_r', GHz	Bandwidth, GHz	S11, dB	VSWR	Gain, dB	Radiation Efficiency, %
1.	4.32	0.7	-23.90	1.10	3.99	90
2.	8.86	0.48	-17.32	1.43	4.16	68
3.	10.10	0.96	-23.02	1.62	7.91	93
4.	12.15	0.97	-13.29	1.79	6.65	84

4. Parametric Analysis

4.1 Performance evaluation with different feedwidths

In the Sub-Section, performance parameters are examined carefully by taking five different feedwidth values. The value of feedwidth chosen are 2.6, 2.7, 2.8, 2.9, and 3.0 mm respectively. Correponding to each feedwidth, the value of 'F_C', S11, bandwidth, VSWR and gain are noted down. All the values are noted down and compared. Antenna with 'F_W'=2.6 mm, shows two operating bands. Three bands are observed with 'F_W'=2.7 mm. Better S11 charactertics are examined with 'F_W'=2.9 mm. Triple bands with suitable gain is evaluated at 'F_W'=3.0 mm. By comparing all the values, best results are noted with 'F_W'=2.8 mm. Table 3 shows the simulated outcomes of the modified multiband elliptical-shaped patch antenna antenna with different feedwidths.

Table 3 Simulated outcomes of the modified multiband elliptical-shaped patch antenna antenna with different feedwidths

Feedwidth 'F_W', mm	'F_C', GHz	S11, dB	Bandwidth, MHz	VSWR	Gain, dB
2.6	3.36	-12.85	0.46	1.25	2.58
	7.25	-29.15	0.85	1.84	4.79
2.7	6.14	-16.21	0.49	1.65	5.47
	8.45	-20.12	0.58	1.49	3.54
	10.58	-22.25	0.75	1.58	2.48
2.8	4.32	-23.90	0.7	1.10	3.99
	8.86	-17.32	0.48	1.43	4.16
	10.10	-23.02	0.96	1.62	7.91
	12.15	-13.29	0.97	1.79	6.65
2.9	3.48	-	0.45	1.25	1.52

		19.48			
	5.54	-26.42	0.74	1.46	3.84
	7.49	-18.12	0.27	1.69	4.46
	9.48	-19.48	0.69	1.48	5.13
3.0	5.49	-28.14	0.49	1.25	3.15
	9.48	-15.45	0.27	1.49	4.46
	11.25	-19.25	0.31	1.66	2.13

4.2 Performance evaluation with different ground plane length values

In the Sub-Section, performance parameters are noted obtained by taking different ground plane length 'L_G'. The value of 'L_G'chosen are:16.5, 17.5, and 18.5 mm, respectively. All the values are compared and tabulated in Table 4. By comparing, it is found that the optimal value of 'L_G' is 17.5 mm. This value is used for designing the antenna structure and is responsible for multiband characteristics.

Table 4 Simulated outcomes of the modified multiband elliptical-shaped patch antenna with different ground plane length values

Ground plane length 'L_G', mm	'F_C', GHz	S11, dB	Bandwidth, MHz	VSWR	Gain, dB
16.5	2.48	-15.48	0.45	1.49	2.48
	6.48	-24.15	0.31	1.57	3.48
	7.49	-16.48	0.42	1.61	5.55
17.5	4.32	-23.90	0.7	1.10	3.99
	8.86	-	0.48	1.43	4.16

		17.32			
	10.10	-23.02	0.96	1.62	7.91
	12.15	-13.29	0.97	1.79	6.65
	6.18	-14.18	0.25	1.25	6.45
18.5	7.19	-18.26	0.82	1.64	4.42
	8.34	-21.36	0.93	1.49	-2.05

4.3 Performance evaluation with different feeding method

In the Sub-Section, performance is noted down with different feeding method. Two feeding techniques are used: microstrip line and coaxial probe. In both cases, the whole structure remains same but difference is in feeding type. All the results are compared and evaluated best feeding method. Table 5 compares the results of the antenna with different feeding method.

Table 5 Simulated outcomes of the parameters of the modified multiband elliptical-shaped patch antenna with different feeding method

Feeding method	'Fc', GHz	S11, dB	Bandwidth, MHz	VSWR	Gain, dB
Coaxial probe	4.52	-13.49	0.42	1.52	3.42
	6.85	-19.54	0.16	1.46	4.49
Microstrip line	4.32	-23.90	0.7	1.10	3.99
	8.86	-17.32	0.48	1.43	4.16
	10.10	-23.02	0.96	1.62	7.91
	12.15	-13.29	0.97	1.79	6.65

5. Comparison with few published antennas

A comparison of the suggested antenna design with published work is revealed in Table 6. Factors considered for comparison of the antennas are material type/ area, working frequencies and gain.

Table 6 Comparison of proposed modified multiband elliptical-shaped patch antenna with few published antennas

Ref. no.	Material/ Area (mm^2)	Working frequencies (GHz)	Gain
[7]	FR4/ 120 x 87	0.36/1.32/5.50	1.91/3.72/7.52 dB
[8]	FR4/ 59 x 90	2.17/4.47/5.6	2/3/2 dBi
[9]	FR4/ 56 x 59	1.57/2.66/3.63	1.63/2.59/3.23 dB
[20]	FR4/ 70 x 70	2.45/5.51	3.90/5.95 dBi
[21]	FR4/ 50 x 60	0.9/2.4/3.8/5.1	0.5/1.125/2.26/ -8.62 dB
[22]	FR4/ 32 x 40	1.7/2.4/3.1/5.4	1.6/2.15/2.75/3.8 dB
[23]	FR4/ 88 x 108	2.0/3.5/4.9/6.5	3.23/4.3/5.95/6.5 dB
Proposed antenna	**FR4/37.5 x 37.5**	**4.32/8.86/10.10/12.15**	**3.99/4.16/7.91/6.65 dB**

6. Conclusion

Wireless technology is preferred because of its important characteristics such as greater mobility and suitable convenience. In this manuscript, FR4 based quad band modified elliptical-shaped microstrip antenna is implemented. For the realized structure, partial ground plane is used. A 50Ω feed line acts as a feeding source to the modified elliptical-shaped patch antenna. It is examined from the results that the resonances occur at 4.32, 8.86, 10.10, and 12.15 GHz with an associated S11 value -23.90,-17.32, -23.02, and -13.29 dB. The gain values evaluated at the claimed frequencies are 3.99, 4.16, 7.91, and 6.65 dB. Radiation patterns are arbitrary bi-directional/ omni-directional.

References

[1] Chowdhury, B. B., De, R., and Bhowmi, M. (Feb, 2016). A Novel Design for Circular Patch Fractal Antenna for Multiband Applications. *3rd International Conference on Signal Processing and Integrated Networks (SPIN-2016),* Noida, pp. 449-453.
[2] Jo, S., Choi, H., Shin B., Oh, S., and Lee, J. (2014). A CPW-Fed Rectangular Ring Monopole Antenna for WLAN Applications. *International Journal of Antennas and Propagation,* **2014**(1), 1-6.
[3] Rahimi, M., Keshtkar, A., Zarrabi, F. B., and Ahmadian, R. (2015). Design of compact patch antenna based on zeroth-order resonator for wireless and GSM applications with dual polarization. *International Journal of Electronics and Communications,* **69**(1), 163-168.
[4] Bhatia, S. S., and Sivia, J. S. (2016). A Novel Design of Circular Monopole Antenna for Wireless Applications. *Wireless Personal Communications,* 1153-1161.
[5] Shakib, M. N., Moghavvemi, M., and Mahadi., W. N. L. (2014). Design of a Compact Tuning Fork-Shaped Notched Ultrawideband Antenna for Wireless Communication Application. *The Scientific World Journal, 2014.*
[6] Singh, J., and Sharma, N. (2016). A Comparison of Minkowski, Compact Multiband and Microstrip Fractals with Meander Fractals Antenna. *International Journal of Computer Applications,* **154**(4), 23-25.
[7] Weng. W. C., and Hung, C. L. (2014). An H-Fractal Antenna for Multiband Applications. *IEEE Antennas and Wireless Propagation Letters,* **13**, 1705-1708.
[8] Mahatthanajatuphat, C., Saleekaw, S., and Akkaraekthalin P. (2009). A Rhombic patch monopole antenna with modified Minkowski fractal geometry for UMTS, WLAN and mobile WiMAX application. *Progress in Electromagnetics Research (PIER,* **89**, 57-74.
[9] Ullah, S., Faisal, F., Ahmad, A., Ali, U., Tahir, F. A., and Flint., J. A. (2017). Design and analysis of a novel tri-band flower-shaped planar antenna for GPS and WiMAX applications. *Journal of Electromagnetic Waves and Applications,* **31**(9), 927-940.
[10] Kaur, K., and Sivia, J. S. (2017). A Compact Hybrid Multiband Antenna for Wireless Applications. *Wireless Personal Communications,* **97**(4), 5917-5927.
[11] Sharma, N., and Bhatia, S. S. (2019). "Performance enhancement of nested hexagonal ring-shaped compact multiband integrated wideband fractal antennas for wireless applications. *International*

Journal of RF and Microwave Computer-Aided Engineering, **30**(3), 1-21.
[12] Kumar, R., Malathi, P., and Sawant, K. (2011). On the design of wheel-shaped fractal antenna. *Microwave and Optical Technology Letters,* **53**(1), 155-158.
[13] Bangi, I. K., and Sivia, J. S. (2018). Minkowski and Hilbert Curves Based Hybrid Fractal Antenna for Wireless Applications. *International Journal of Electronics and Communications,* **85**, 159-168.
[14] Sundaram, A., Maddela, M., and Ramadoss, R. (2007). Koch-Fractal Folded-Slot Antenna Characteristics. *IEEE Antennas and Wireless Propagation Letters,* **6**, 219-222.
[15] Nasimuddin, Chen, Z. N., and Qing, X. (2013). Slotted Microstrip Antennas for Circular Polarization with Compact Size. *IEEE Antennas and Propagation Magazine,* **55**(2), 124-137.
[16] Murugan, S., and Rajamani, V. (2012). Design of Wideband Circularly Polarized Capacitive fed Microstrip Antenna. *Procedia Engineering,* **30**, 372-379.
[17] Rafi, Gh. Z., and Shafai, L. (2004). Wideband V-slotted diamond-shaped microstrip patch antenna. *Electronics Letters,* **40**(19).
[18] Tanweer, A., Prasad, D., and Rajashekhar. B. (2018). A miniaturized slotted multiband antenna for wireless applications. *Journal of Computational Electronics.*
[19] Sree, G. N. J., and Nelaturi, S. (2021). Design and experimental verification of fractal based MIMO antenna for low sub 6-GHz 5G applications. *International Journal of Electronics and Communications,* **137**(10).
[20] Kumar, S. A., and Dileepan, D. (2017). Design and development of CPW fed monopole antenna at 2.45 GHz and 5.5 GHz for wireless applications. *Alexandria Engineering Journal,* **56**(2), 231-234.
[21] Karia, D. C., Goswami, S. A., and Dhengale, B. (2015). A compact wideband antenna for wireless applications using rectangular ring", 1st International Conference on Next Generation Computing Technologies (NGCT-2015). *1st International Conference on Next Generation Computing Technologies (NGCT-2015),* Dehradun, pp. 560-562.
[22] Mark, R., Mishra, N., Mandal, K., Sarkar, P.P., and Das, S. (2018). Hexagonal ring fractal antenna with dunb bell shaped defected ground structure for multiband wireless applications. *International Journal of Electronics and Communications,* **94**, 42-50.
[23] Gupta, A., Joshi, H. D., and Khanna, R. (2017). An X-shaped fractal antenna with DGS for multiband applications. *International Journal of Microwave and Wireless Technologies,* **9**(5), 1075-1083.

[24] Kaur, M., and Sivia, J. S. (2019). Minkowski, Giuseppe Peano and Koch Curves based Design of Compact Hybrid Fractal Antenna for Biomedical Applications using ANN and PSO. *International Journal of Electronics and Communications*, *99*, 14-24.
[25] Balanis, C. A. *Antenna theory: Analysis and design.* (3. Edition, Ed.) London: John Wiley & Sons.
[26] Kaur, M., and Sivia, J. S. (2019). Giuseppe Peano and Cantor set fractals based miniaturized hybrid fractal antenna for biomedical applications using artificial neural network and firefly algorithm. *International Journal of RF and Microwave Computer-Aided Engineering,* **30**(1), 1-11.
[27] Kaur, M., and Sivia, J. S. (2020). ANN and FA Based Design of Hybrid Fractal Antenna for ISM Band Applications. *Progress in Electromagnetics Research C,* **98**, 127-140.
[28] Kaur, M., and Sivia, J. S. (2019). ANN-based Design of Hybrid Fractal Antenna for Biomedical Applications. *International Journal of Electronics,* **106**(8), 1184-1199.
[29] Gupta, M., and Mathur, V. (2017). Wheel shaped modified fractal antenna realization for wireless commucations. *International Journal of Electronics and Communication,* **79**, 257-266.
[30] Dib, N. I. (2015). Synthesis of thinned planer antenna arrays using teaching-learning-based optimization. *International Journal of Microwave and Wireless Technologies,* **7**(5), 557-563.
[31] Gill, H. S., Khera, B. S., Singh, A. and Kaur, L. (2019). Teaching-learning-based optimization algorithm to minimize cross entropy for Selecting multilevel threshold values. *Egyptian Informatics Journa*, **20**(1), 11-25.
[32] Rao, R. V., and Patel, V. (2013). An improved teaching-learning-based optimization algorithm for solving unconstrained optimization problems. *Scientia Iranica,* **20**(3), 710-720.
[33] Zadehparizi, F., and Jam, S. (2019). A new chaotic teaching learning based optimization for frequency reconfigurable antennas design. *Journal of Intelligent and Fuzzy System,* **36**(1), 1-8.
[34] Rao, R. V., and Waghmare,G. G. (2014). A comparative study of a teaching-learning-based optimization algorithm on multi-objective unconstrained and constrained functions. *Journal of King Saud University-Computer and Information Sciences*, **26**(3), 332-346.
[35] Ahmad, A., Arshad, F., Naqvi, S. I., Amin, Y., Tenhunen, H., and Loo, J. (2018). Flexible and Compact Spiral-Shaped Frequency Reconfigurable Antenna for Wireless Applications. *IETE Journal of Research*, **66**(1), 22-29.

[36] Varamini, G., Keshtkar, A., Daryasafar, N., and Moghadasi, M. N. (2018). Microstrip Sierpinski fractal carpet for slot antenna with metamaterial loads for dual-band for wireless applications. *International Journal of Electronics and Communications*, **84**, 93-99.

Chapter 5

Exploring Traditional versus Online Learning Model of Computer Science in North India

Inderpreet Kaur[1] and Kiranjeet Kaur[2]
[1,2] Department of Computer Science
[1]Mata Sahib Kaur Girls College Talwandi Sabo
[2]Guru Nanak College Moga

1. Introduction

Since February 2020 Covid-19 is incessantly disseminating mainly the education sector is the worst hit so the blended mode is the best alternative to take hold of this situation. During this pandemic, e-learning has become the most popular and the fastest-growing "business" and every week new online courses are launched for universities, institutions or schools. Online education has become more familiar in the modern era. Online courses are useful for all students having different background even if they are from diverse regions, states and even from diverse countries.

The concept of not regular (part-time, correspondence) courses started in universities to help educate aspiring students who cannot pursue regular courses for different reasons, maybe distance, scarcity of time or other social reasons. The leading universities that started courses through distance learning were from south of India viz Anna University, Annamalai University, Madurai Kamraj University. It is a very sensitive task to provide education through education material to the students enrolled from remote places. Some universities also started delivering lectures through radio broadcastings in the 1980s.

Another concept was also started of Open Universities, whereas there was no need of prior education before joining the offered courses as pre-requisites. In north, the Rajasthan state's Kota Open University was the first one to start with the Open University concept. The main reason behind the idea was to make the state population literate as the literacy level of this state was very low.

The computerisation of banks and industry opened up avenues for job seekers. Obviously, the demand for skilled computer operating staff started increasing. Thus the need for certified courses arose. The trend of lecture broadcasting changed from radio to television with the introduction of Indra Gandhi National Open University in New Delhi. Various courses were started catering to different streams from Vocational, Arts, Science and Engineering. Herein we are most concerned about the courses for computer literacy.

The journey of learning in Information Technology (or Computer Science/Application) starts from traditionally closed to open, online learning and has spread its advances in research also. By closed learning means the traditional courses started; where the learner joins course with fixed hours and years to complete it successfully to achieve a certificate or degree at a fixed location namely school/college/university. Online learning has too much flexibility to complete the course without any bindings of attendance and timing schedules. Online learning courses have huge benefits but have some disadvantages also. (See Table 1)

Table 1: Merits and Demerits of Traditional v/s Online Education

Traditional learning	Online Learning
Merits 1. There is Face to face or interaction between teachers and learners 2. This kind of interaction is too easy to conduct campus activities 3. This is the appropriate way to give training to the students individually.	**Merits** 1. More tractability for those who work for the whole time 2. Online Education enables the students to complete projects at times when it is most suitable 3. It can acquire a degree from any institution without moving 4. It is impeccable for military students
Demerits 1. Rigid in class schedules 2. Economical 3. Time-consuming and more	**Demerits** 1. Indirect contact between learners and instructors 2. It provides very few oppor-

travel to and from class	tunities for campus events 3. Sometimes "technical difficulties" and software problems

There are two ways of online learning: **synchronous** and **asynchronous**. Synchronous learning is instruction and engages students in learning in "real-time" via the Internet. It usually involves tools, such as:

- Online chat
- Audial and Video conferencing
- Data and Application allocation
- Shared whiteboard
- "Hand raising" Indicator
- Combined programming of multimedia presentations and Live slide shows

Asynchronous learning uses time-delayed proficiencies of the Internet. It consists of many tools such as:

- E-mail
- Threaded discussion
- Newsgroups and bulletin boards
- File attachments

Asynchronous courses are still instructor-assisted but are not managed in real-time, which means that students and teachers may involve in activities related to their courses as per their appropriateness rather than during exactly in time class sessions. But asynchronous courses, learning does not need to be programmed like synchronous learning. It allows benefit to students and mentors at any time, anywhere.

2. Online learning modules

At the international level, Open Software Systems is very popular for a long time, the Indian education system was also adopted according to the suitability of the education aspirants but it is moving at a very slow pace. But the growth in the number of students is seen since the burgeoning rise in internet users mainly on smartphones and other handheld devices. We can take the case of NPTEL, SWAYAM (Study Web of Active Learning by Young and Aspiring Minds) and

IIT spoken tutorials the pioneer in delivering open courses suitable for Indian students.

The NPTEL developed various video courses based on syllabus (110 new courses, 109 present courses condensed in digital visual format and 129 web-based e-courses) and accepted by 7 (seven) IITs, IISc Bangalore as Partner Institutions (PI) and various additional esteemed institutes as Associate Partner Institutions (API) through combined efforts. The primary motive of NPTEL is to frame a specific and the best learning content offered for the students of engineering institutions all over the country by discovering the high-quality technologies in ICT. It was the first time in which all IITs and IIMs have a common aim and proposed for working jointly to expand the level of science, engineering and management education all over the regions by providing modules through VCTEL.

SWAYAM (Study Web of Active Learning by Young and Aspiring Minds) is a programme vigorously started by the Regime of India and is designed to attain the three fundamental principles of Education Plan: Access, Equity and Quality. The main purpose is to bring leading teaching-learning sources. Courses which are delivered through SWAYAM, free for the learners, and the certificates provided to the students are registered and presented a certificate on successful completion of the course having very less fee. Courses of Computer Sciences are offered by Public Portal (NPTEL, Swayam and IIT Spoken Tutorial) like Introduction to Internet of Things, Software Engineering, C, C++, Cloud Computing, Data Base Management System etc.

One more initiative has been taken by The Government of India. But it needs help from the country's leading institutes to take participate in the DIGITAL INDIA project and promote the use of FOSS (Free Open Source Software) to achieve the objectives explained before. The IITs have started spoken tutorials concept to access students and institutions to promote FOSS. It became successful in acquiring the desired goals. The Ministry of Human Resources and Development, Government of India launched The Spoken Tutorial project (on the 26th of January, 2010) of the 'Talk to Teacher' activity of the National Mission on Education through Information and Communication Technology (ICT). Learning becomes more effective when animation and narration are offered all the way together.

The IIT spoken tutorials started their operations in 2011. The total number of training/workshops conducted since 2011 is 68729, for 3514891 participants in 3770 institutions from all over India. The learning model of IIT spoken tutorials is unique in the way that the students enrol at the institution they get admission in the regular course and complete the tutorials under the supervision of a local trainer and get online help too. This is an advantage that they do not need to spare extra time for practice at home/hostel after regular classes and another advantage is that the courses offered sync with the regular course they are supposed to clear in a particular year/semester. Hence, this way the students and the institute both get an advantage of using free of cost software for conducting practical examinations and no cost of hardware licensing, online training support etc. and the IIT spoken tutorials does not need to appoint any supervisor at the learning location. The enrolled student is under master trainers supervision online at the central location. [http://spoken-tutorial.org/]

The main goals of these projects are to bring in all the best teachers in the country within the sphere of online learning and record their lectures/seek their collaboration with IITs/IISc and make their courses accessible for the public under free and open sources agreement.

3. Literature Survey

Rajpal et al. (2008) concluded that with the advent of technologies, a turbulent change is occurring in Indian education. As India is facing many challenges in education and in training, it becomes essential to use e-learning which provides the answers to them. Authors discussed the three case examples such as Indian Institute of Management Ahmadabad (IIM-A), Symbiosis Centre for Health Care (SCHC) and Amrita. Authors concluded that E-learning models can be established and executed by different institution of higher education from corner to corner in the country as the majority of the population is interested in education. It needs to be addressed by the organizers, creators and entrepreneurs for the advantage of nations.

Anderson and Gronlund (2009) discussed that there are four kinds of challenges in the operating and achievement of e-learning. These challenges are individual challenges which included motivation, contradictory priorities, economy, academic confidence, gender and age. The second challenge is course challenges which consisted of design, curriculum, academic models, content, teaching, learning activities and flexibility. The third challenge is contextual challenge referred to role of teachers and students, the attitude of e-learning/IT, training of teachers as well as staffs along with set of rules in addition to regulations. Technological challenges are technological requirements such as radio, computer, cassettes and so on as well as the main factor is the cost of these technologies. Thus, all challenges are necessary to bring the changes in education and to enables e-learning effective.

Dighe (2010) examined that ICT is a tool to improve literacy innovation. The area which is neglected by literacy had non-availability of experts and inactive involvement for technical maintenance. The EGYPT country is one of them who are not much users of ICT. As ICT has become an important strategy that support the delivery of basic skill in adult education. It is used as a powerful technique to raise the standard of literacy and numeracy. It also provide attractive way of learning by using computer, multimedia and web. By ICT, adult education can become efficient, interactive, self-paced and flexible. ICTs and media such as newspapers, radio, T.V, computers and so on brought some transformation in the quality of education as well as knowledge of adult learners.

Siddiqui and Masud (2012) discussed about E-learning. It was the utilization of existing information and communication technologies to enable the learning process. In the global world learners are looking for an innovative way of learning and knowledge with ease and internet. E-learning is the way in which students can fulfil their needs. E-leaning is the best way to improve the higher-level education system and also helps to deliver the lectures and contents in remote areas. Web application via mobile ensures benefits to the students.

Iqbal and Islam (2013) concluded in this study that e-learning and knowledge management enhance an individual's and organization's knowledge skill. But knowledge management is more fresh and current activity as compared to an e-learning course. E-learning has become outdated now. Thus, e-learning and KM both need to be together by feeding the content into e-learning to make it more fresh and learners were to be tapped into a sustainable knowledge after completing the course as well as e-learning should be feed to KM by providing an easy method for organizing information. If by converging these two technologies, great impact in the learning process was examined.

Upasana (2014) discussed traditional education as well as e-education in India. Traditional education is one-way communication by the teacher to the learners. It is more beneficial to those students who want to interact face to face with the gurus. Traditional education provides knowledge, skills and it is reasonable for every class of people. It enables the learner fruitful for their own and other people's welfare. On the other side, e-learning refers to teaching and learning for all levels of people by using networking and communications technology but it acquires a huge amount of infrastructure facilities. Both students and the teachers must have a least possible knowledge of computer to work effectively in an online atmosphere. The learning can be done at our own pace and schedule through e-learning as well as it requires writing skills, ability to navigate, create information using digital technologies, more self-direction and discipline as compared to traditional education.

Singh and Kaur (2015) observed that e-learning is not fresh but it has developed like WWW in every country. E-education means to give education and training via internet. India consists of different students of different socio-background and different economic conditions. It is very difficult to change these aspects but we can provide the web-based learning that is uniform teaching-learning resource. To adopt e-learning, higher education institutions have to face many challenges. Despite these challenges, e-learning allows interactive content delivery which overcomes the restrictions of traditional resources. It provides flexible environment for the students to inculcate the habit of self-study. To popularize e-learning, various organizations such as IGNOU, NCERT, EKALAVYA, NPTEL and so on are taking many initiatives by providing e-textbooks, e-content as well as interactive animations to teach high-quality learning material. Students of every field can get the advantage of e-learning by staying at home.

Msomi (2016) concluded that e-learning in the educational system benefits the universities and higher education institutes. There is paradigm shift in the teaching and learning process. There is the use of innovative methods in the teaching and learning process. The methods and materials used in e-learning meet the quality of teaching as well as the learning process. Revision and updating the contents should be an easy process for the learners. Faculty and student's satisfaction are the main factors for the success of e-learning.

Bhatt and Maniar (2017) have observed that the institutions are always at loss and lacking of knowledge who do not realize the importance of E-learning. Nowadays E-learning has become the essential need of modern education in Indian colleges and universities to enter in the international standards as well as to attract affordable foreign students. So the subjects, courses must be settled in such a way so they may fulfil the international standards and needs of every student. In higher education, a teacher-free classroom has become more popular. The global world of knowledge is possible if it is through technology-based learning.

Michal Baczek, Michalina Zaganczyk, Monika Szpringer, Andrzej Jaroszynski, Beata Wo (2021) observed has that e-learning played a very major role in the pandemic situation of Covid-19. It is very effective to enhance not only knowledge but also improve social skills. The majority of the students accepted this mode of learning in such situation. e-learning become an active approach where the student works with material and even get feedback at the same time.

4. Objective

The prime objective of this research is to examine the effectiveness of online learning platforms on the ability of subscribers to improve employability as well as knowledge. The field of online learning is ironic in studies focused on evaluation issues, particularly from the students' perspective. This research also emphasises the place of residence, level of education and age, which influence the subscribers and examine the quality of service and other OLLP features. This research focuses on analysing the differences in perceptions of online classes between students who want to take online classes and those who do not want to take online classes. The stream of this research paper is:

a) Section I provides the Introduction and literature survey.

b) Section II analyses the results and discussions.
c. Section III summarises the paper in Conclusion.

5. Research Questions

The success of IIT spoken tutorials, NPTEL and SWAYAM online campaign shows table-thumping success with large number of enrolments. It is obvious that the students already enrolled in colleges are opting for online lessons and online examination system along with the traditional education delivery system. How does this impact the education delivery method is the area of concern here.

There are certain research questions that need to be answered.

1. Why do online education delivery gain popularity over traditional education delivery methods and the level of effectiveness of online courses described by high levels of different types of interactions?
2. Which online course design (i.e., emphasize learner-content, learner-teacher or learner-learner interconnections or interactivity) results in the highest levels of efficacy?
3. What features make online courses for computer courses popular amongst learners in short time compared with other online education delivery model(s)?
4. What is the future scope of this model on education delivery methods on other subject(s)?

6. Research Methodology

Two different research methods are used in this research. A qualitative study was used for the very first phase where specific attributes were identified and confirmed through review of literature and analysis of content. The content analysis comprised research of OLLP system services that contained particular of their OLLP system projects available for students viewing and its implementation or usage. OLLP system for various northern region states of India was studied and extracted a list of attributes. The second phase is used for output from the first phase to develop a comprehensible Questionnaire that was forwarded to collect responses of students of Haryana, Himachal Pradesh, Jammu, Punjab and cities like Chandigarh and Delhi. Graduate and Postgraduate students were asked to rate their view on the attributes in the context of OLLP system. The primary motive of this survey was to gain qualitative data to conclude which attributes should be counted in OLLP system, thereby answering the research questions. A well-defined questionnaire was passed among people of various degree colleges, engineering colleges and universities, containing both graduate as well as postgraduate students, to have a better view regarding the OLLP system implementation. Total of 277 students of northern region of India having different educational background like Graduate/Post-Graduate and of different age participated in this survey. There are eleven items (A-K) in this questionnaire to measure the quality, features and performance characteristics of OLLP system. The abbreviations (A-K) are being used for the items of the questionnaires in Table 2.

Table 2 Questionnaire Part -1

A	OLLP can be retrieved without any technical boundaries (hardware/ software).
B	OLLP is too easy to run.
C	OLLP is helpful to understand the subject in concern.
D	OLLP offers data/ text/ audio/ video/ power-points formats.
E	OLLP delivers hyperlinks to outside resources.
F	OLLP holds an interactive chat system (chatbot) to assist.
G	OLLP is easy to use.
H	OLLP sought out subscriber's problems fast.
I	OLLP is designed as per student's concern.
J	OLLP is amended regularly as per student's needs.
K	OLLP system give emphasizes on subjects trained in particular university only.

Table 3: Correlation coefficient (OLLP) student satisfaction

	#1	#2	#3	#4	#5	#6	#7	#8	#9	#10	#11
A	1.00										
B	0.78	1.00									
C	0.65	0.85	1.00								
D	1.00	0.78	0.65	1.00							
E	0.22	0.02	0.14	0.22	1.00						
F	0.24	0.28	0.38	0.24	0.27	1.00					
G	0.70	0.91	0.77	0.70	0.09	0.21	1.00				
H	0.38	0.38	0.47	0.38	0.08	0.23	0.31	1.00			
I	0.41	0.09	0.03	0.41	0.18	-0.05	0.05	0.44	1.00		
J	0.20	0.47	0.36	0.20	0.25	0.02	0.66	0.13	0.00	1.00	
K	-0.24	-0.21	-0.15	-0.24	0.04	-0.26	-0.33	-0.15	0.09	-0.02	1.00

The correlation between item 1 and item 2 is 0.78, item 1 and item 3 is 0.65 and so on. As Cronbach's alpha is calculated, it comes to 0.84. It is quite high. It is considered reliable. Even though the correlation coefficient is low in some of the individual items. The number of items (eleven) in this question is sufficient as suggested by statistical analysts. It is also obvious that as the number of items increases the reliability is expected to increase too.

Table 4: Cronbach's alpha of OLLP parameters for student's satisfaction
(Reliability, if a term is dropped)

	raw_alpha	std.alpha	G6(smc)	average_r	S/N	alpha se	var.r	med.r
A	0.720	**0.80**	0.90	0.29	4.00	0.061	0.064	0.24
B	0.710	**0.80**	0.90	0.29	4.00	0.063	0.058	0.24
C	0.710	**0.81**	0.91	0.29	4.10	0.064	0.066	0.24
D	0.720	**0.80**	0.90	0.29	4.00	0.061	0.064	0.24

E	0.770	**0.85**	0.93	0.36	5.60	0.051	0.076	0.31
F	0.750	**0.84**	0.93	0.35	5.30	0.056	0.080	0.25
G	0.710	**0.80**	0.89	0.29	4.00	0.065	0.061	0.24
H	0.730	**0.83**	0.92	0.33	4.90	0.059	0.083	0.24
I	0.770	**0.85**	0.93	0.36	5.60	0.052	0.070	0.26
J	0.750	**0.84**	0.92	0.34	5.20	0.055	0.075	0.26
K	0.800	**0.85**	0.93	0.36	5.60	0.045	0.075	0.28

The value of correlation is less than 0.36, this indicates that the item may not belong on the scale. The reason for such a low value may be because some items of the questionnaire depend on each other and some do not. Thus, we consider Cronbach alpha as reliability measure only. The acceptable value of standard Cronbach alpha that is commonly acceptable is 0.70, here in this case. The coefficient for these items is 0.84 at 95% level of confidence. The reliability of the data when an item is dropped is shown in the Table. The coefficient remains between 0.80 and 0.85 if any item is dropped. Thus, there is no need to drop any item from the question to increase the reliability of the data. In another question, the quality of the online education portal is measured (in 5point, Likert scale) in terms of efficient delivery of services to help respondents sharpen their skills.

Table 5 Questionnaire Part –II - There are seventeen items in this questionnaire of part-II that measure the OLLP facilitates the efficient delivery of effective skill services to assist students in accomplishing additional qualifications and meet up with the student's interest.

Table 5: Questionnaire Part -II

A	Facility of OLLP is framed to suit online use.
B	The course goes very smoothly all over the program.
C	In OLLP, the service is presented to the students directly in a precise and easy language.
D	The services of OLLP give particular consideration to the student requirement.
E	The services of OLLP are new and updated.
F	Students are fully satisfied with the services and courses of OLLP system.

G	The OLLP provides the intensive/ complex doubts of the subject
H	Course content/ Assignments are streamlined time to time.
I	The contents/ Tests / Tasks are reformed on student's demands.
J	Course content is easy to access and adds to the information.
K	OLLP provides courses as per the university curriculum.
L	The concepts/ skills are useful and meet the academic and industrial demands.
M	Enhancements of skill concepts are productive and beneficial.
N	The concepts/ skills of the course content are comprehensive and enough for employability.
O	The course concepts/ skills are uploaded on basis of the schedule.
P	The complete course is uploaded in one go and is referred well-timed.
Q	The course concepts/ skills are according to an institution of higher education in India.

Table 6: Cronbach's alpha of OLLP parameters for efficient delivery of services (Reliability, if an item is dropped)

	raw alpha	std. alpha	G6(smc)	average_r	S/N	alpha se	var. r	med. r
A	0.82	**0.82**	0.93	0.22	4.50	0.04	0.05	0.20
B	0.79	**0.80**	1.00	0.20	4.00	0.04	0.05	0.18
C	0.79	**0.79**	0.91	0.19	3.80	0.04	0.04	0.18
D	0.81	**0.81**	0.92	0.21	4.40	0.04	0.05	0.21
E	0.79	**0.80**	0.91	0.20	3.90	0.04	0.05	0.18
F	0.81	**0.82**	0.93	0.22	4.50	0.04	0.05	0.19
G	0.82	**0.83**	0.93	0.23	4.70	0.04	0.05	0.22
H	0.81	**0.82**	0.93	0.22	4.40	0.04	0.05	0.21
I	0.81	**0.81**	0.92	0.21	4.40	0.04	0.05	0.20
J	0.81	**0.80**	0.91	0.20	4.10	0.04	0.04	0.21
K	0.82	**0.83**	0.92	0.23	4.8	0.04	0.05	0.22

					0			
L	0.80	**0.80**	0.92	0.20	4.10	0.04	0.05	0.19
M	0.81	**0.80**	0.91	0.20	4.10	0.04	0.04	0.21
N	0.81	**0.82**	0.92	0.22	4.40	0.04	0.05	0.20
O	0.82	**0.83**	0.93	0.24	5.00	0.04	0.05	0.23
P	0.81	**0.81**	0.91	0.22	4.40	0.04	0.05	0.21
Q	0.82	**0.83**	0.93	0.23	4.70	0.04	0.05	0.22

The coefficient for these items is 0.82 at 95% level of confidence. The reliability of the data when an item is dropped is shown in the Table. The coefficient remains between 0.79 and 0.83 if any item is dropped. Thus, there is no need to drop any item from the question to increase the reliability of the data.

ANOVA test is applied to discover if there is any significant effect of Age, Place of Residence (Location) and the Qualification of the respondent using OLLP for services, content and quality parameters on five-point Likert scale as follows:

- OLLP system is convenient.
- OLLP system helps thorough understanding of subject in concern.
- In OLLP, the service is offered to the learner right away in an understandable language.
- The concepts and skills of the course material are comprehensive and appropriate for employability.
- Certificate provides Satisfaction to the students.

7. Results and Discussion

The data is collected through questionnaire online and offline mode. The location-wise, age-wise and qualification-wise data is collected to find out the experiences about the services of the portal from presentation and content point of view. The summary of the analysis is being reproduced as follows:

- It was observed that there is a variance of the opinion of respondents from different age groups about the operability. It was noticed that young generations feel easy to operate online portals and understands topics through an electronic medium more easily than the elder ones.

- The study also reveals that there is a difference of thoughts of respondents from different age groups about providing the content of the online learning portal being sufficient enough for employability and about providing the certificate after attending the online learning portal course.

- Location wise respondents are more concerned with the issuance of certificates after completion of the course for employability than others and the young generation feels more concerned to understand the content of the course for employability and with the provision of clear and simple language from online learning portal than the elder ones.

- It was also analyzed that there is a difference of views of respondents from different age groups that the young generation feels more concerned to the issuance of certificate after completion of the course for employability than the elder ones.

- It has been observed that there is a contrast of opinion of respondents from different locations about the operability of the online learning portal. Respondents of big cities feel easy to operate online portals and are more concerned with understanding the topic through an electronic medium easily than others.

- There is a dissimilarity of opinion of respondents with different educational qualification levels about the operability of the online learning portal and about the performance of the online learning portal in an understanding subject in concern. Respondents from big cities feel easy to operate online portals more easily than others.

- Study reveals that there is a difference of opinion of respondents with various qualification levels about providing the content of the online learning portal using clear and simple language and is sufficient enough for employability. It may be because of location, cultural disparity, level of literacy and understanding level.

- The majority of the respondents were agreed with the provision of clear and simple language used by the online learning portal than the others.

- There is a difference of opinion of respondents having different qualification levels belonging to different places of residence about providing the certificate after attending an online learning portal course. Location-wise respondents are more concerned about the issuance of certificates after completion of the course for employability than others.

8. Conclusion and Future scope

To deal with enrollment, many institutions during the Covid-19 pandemic are shifting to online education because it was the need of that time so the popularity of online classes has increased. It is necessary to understand whether online education is effective and which factors are responsible to make it effective. It is also helpful for tutors to cognize how best to teach and interact with learners in these platforms of online learning. Due to the growing popularity of online platforms, this research has become able to answer about efficiency and viewpoint of students to word strength and weakness of online learning. This can be used when an administrator is deciding to adopt online learning platforms for their institutes. With an increase of new variants of covid-19 schools, colleges and universities are shut down and millions of students were stuck at home. The virus had an austere effect on educational institutions therefore the only way to interact with students is through online learning. Online education is safe and reliable option to ensure the health of teachers and students alike. This research will give direction to the scholars doing research in this domain.

References

1. Allen E. I. and Seaman J. (2013). Changing Course: Ten Years of Tracking Online Education in the United States. Babson Survey Research Group.
2. Chambers, T. E. (2002). Internet course student achievement: In Ohio's two-year Community and technical colleges, are online courses less effective than traditional courses?

3. De Freitas and S. Neumann, T. (2009). The use of 'exploratory learning' for supporting immersive learning. Computers & Education, **52**(2), 343–345.
4. Dickey, M. D. (2003). Teaching in 3D: pedagogical affordances and constraints of 3D virtual worlds for synchronous distance learning. Distance Education, **24**(1), 105–121.
5. Doom, J. R. and Doom, J. D. (2014). The quest for knowledge transfer efficacy: blended teaching, online and in-class, with consideration of learning typologies for non-traditional and traditional students. Frontiers in Psychology, **5**, 324.
6. Estes C. (2013). Online learning is bringing huge change to higher education. What have we learned so far? 95–106.
7. Esteves, M., Fonseca, B., Morgado, L. and Martins, P. (2008). Contextualization of programming learning: a virtual environment study. In Proceedings of the 38th ASEE/IEEE Frontiers in Education Conference, October 22–25, 2008, Saratoga Springs, NY (pp. 17–22). Washington, DC: IEEE.
8. Farinella, J. A., Hobbs, B. K., and Weeks, H. S. (2000). Distance delivery: The faculty perspective. Financial Practice and Education, **10**, 184–194.
9. Feenberg, A. (1998). "The Written World: On the Theory and Practice of Computer Conferencing." In Mason, R. and Kaye A. (Eds), Mindweave: Communication, Computers, and Distance Education. Oxford: Permagon Press.
10. Gomes, A., Areias, C. M., Henriques, J. and Mendes, A. (2008). Aprendizagem de programação de computadores: dificuldades e ferramentas de suporte. Revista Portuguesa De Pedagogia, **42**(2), 161–179.
11. Jenkins, T. (2002). On the difficulty of learning to program. In Proceedings of 3rd Annual LTSN_ICS Conference, Loughborough University, UK, August 27–29, 2002 (pp. 53–58).
12. Kim, K., and Bonk, C. J. (2006). The future of online teaching and learning in higher education: The survey says. Educause Quarterly, **29**(4), 22.
13. Lahtinen, E., Mutka, K. A. and Jarvinen, H. M. (2005). A study of the difficulties of novice programmers. In Proceedings of the 10th Annual SIGSCE Conference on Innovation and Technology in Computer Science Education (ITICSE 2005). Monte da Caparica, Portugal, June 27–29, 2005 (pp. 14–18).
14. Micaela E., Benjamim F., Leonel M. and Paulo M. (2010). Improving teaching and learning of computer programming through the

use of the Second Life virtual world. British Journal of Educational Technology, 1-14. doi:10.1111/j.1467-8535.2010.01056.x

15. Michał Ba̧czek, MD , Michalina Zaganczyk-Ba, czek, MD, Monika Szpringer, MD,PhD, Andrzej Jaroszynski, MD, PhD, Beata Wo, zakowska-Kapłon, MD, PhD (2021) Students' perception of online learning during the COVID-19 pandemic A survey study of Polish medical students, Ba̧czek et al. Medicine (2021) 100:7.
16. Miliszewska, I. and Tan, G. (2007). Befriending computer programming: a proposed approach to teaching introductory programming. Journal of Issues in Informing Science & Information Technology, **4**, 277–289.
17. Motil, J. and Epstein, D. (2000). JJ: a language designed for beginners (less is more). Mya, P., Martha L.A. (2010) Teaching and Learning Online, University of Massachusetts.
18. Olapiriyakul, K. and Scher, J. M. (2006). A guide to establishing hybrid learning courses: employing information technology to create a new learning experience, and a case study. The Internet and Higher Education, **9**, 287–301.
19. O'Kelly, J. and Gibson, J. P. (2006). Robo Code & problem-based learning: a non- prescriptive approach to teaching programming. (Bologna, Italy, June 26–28, 2006) (pp. 217–221). ITICSE '06. New York: ACM.
20. Pape, L. (2010). Blended Teaching & Learning. School Administrator, **67**(4), 16–21.
21. Robins, A., Rountree, J. and Rountreen, N. (2003). Learning and teaching programming: a review and discussion. Computer Science Education, 13, 2, 137–172.
22. Rajpal, S., Singh, S., Bhardwaj, A., and Mittal, A.,(2008) "E-learning Revolution: Status of Educational Programs in India", Proceedings of the International Multi conference of Engineers and Computer Scientists, **1**, 19-21.
23. Schulte, C. and Bennedsen, J. (2006). What do teachers teach in introductory programming? In Proceedings of the Second International Workshop on Computing Education Research, Canterbury, UK, September 9–10, 2006 (pp. 17–28). ICER '06. New york: ACM.
24. Warburton, S. (2009). Second Life in higher education: assessing the potential for and the barriers to deploying virtual worlds in learning and teaching. British Journal of Educational Technology, **40**(3), 414– 426.

25. Winslow, L. E. (1996). Programming pedagogy—a psychological overview. SIGCSE Bulletin, **28**, 17–22.

Chapter 6

Technologies and Solutions for Post Disaster Victim Localization

Todd Murray*, Syed Faraz Hasan** and Morio Fukuoka*
**** Massey University, New Zealand***
***** University of Buraimi, Oman***

1. Introduction

1.1 Background

People inevitably get trapped under rubble after natural and man-made disasters. Because of being trapped, they are oftentimes incapable of calling for help. Natural and man-made disasters, such as earthquakes, tsunamis, floods, and hurricanes, seem to occur regularly without discrimination of borders. In the year 2021 alone, there were a significant number of fatal disasters, such as the magnitude 7.2 earthquake that struck Haiti in August. It is widely agreed that the localization and extrication of trapped victims are most successful within the first 24 hours of such an event occurring [1, 2]. Legacy devices that are used for search and rescue operations include infrared cameras, canines, seismic sensors and so forth. Whilst these devices have proven their worth in the field, they are often difficult to mobilize and time consuming to utilize over a large scale in a timely fashion. This is especially apparent in developing nations. In this context, detecting victims' mobile phones (that are trapped with them) can be a potential solution for rapid localization. Mobile phones are prevalent in today's world, and most people in developed countries own a mobile phone. Over half of the global population now own a smartphone and in higher income countries (i.e. Europe and across North America), smartphone uptake is over 80% [3]. These statistics highlight the importance of developing a system which utilizes such widespread technology.

Mobile phones typically communicate with external infrastructure (GPS, etc.) in order to determine their position and transmit the same to others, if applicable. In a typical rubble environment, communication with external infrastructure may not always be possible. In order to use mobile phone signals for localization, rescuers need to be able to scan the disaster site at particular frequencies to capture wireless signals from a victim's mobile phone using specialized equipment placed nearby.

This chapter explores the extended implementations of cellular based-victim localization systems for use after natural and man-made disasters. The primary focus of this discussion is to examine literature that utilizes cellular technologies to locate victim's mobile phones. This chapter builds from the authors' previous review of literature on a topic that was much broader in domain and focused on the more holistic research around natural disasters and the general technologies used to locate victims [1]. The previous work examined implementations that were primarily immobile (not moving) systems.

In order to warrant the development of a more advanced search and rescue system, research into more complex cellular localization implementation must be performed. This chapter presents a comprehensive discussion around the state of the art of existing search and rescue systems and technologies.

1.2 Current state of the art

Some common tools used in SAR events include visual and infrared cameras, canines, and seismic sensors [4-7]. Whilst most of these tools are effective, most of them are ground-based tools (i.e., seismic sensors, canines). As for cameras, these can be deployed on the ground or in the air (i.e., aboard a helicopter or UAV). Ground-based systems are inherently less mobile than an aerial-based system and are more prone to obstructions. For camera-based systems, there is a distinct trade-off between sensor distance and resolution. Also, any camera-based system is going to be heavily inhibited by visual or thermal obstruction (i.e., building rubble, flooding). Sensors such as seismic sensors are fairly accurate, however they are prone to environmental noise and localisation inaccuracies, particularly in inhomogeneous disaster environments (i.e., building rubble). This is especially apparent if the sensors use sound amplification. Canines are widely regarded as one of the most effective SAR tools, however the canine squad needs to be within close-range of the disaster site.

Additionally, interpretation of canines is difficult for untrained handlers and the general public. With the exception of aerial-based system, many of these tools have a certain working range, and would require multiple deployments in a large-scale disaster. New technologies developed for SAR purposes should address some of these limitations, including mobility, deployability, operating range and should also aim to locate victims rapidly.

The current state of the art of cellular-based search and rescue (SAR) systems is limited. Cellular-based search and rescue tools are on the rise. This growth is predicated on various factors. For example, the work in [8] discusses the candidacy of cellular technology to future integrated SAR systems due to the uptake of mobile phones in modern society. Additionally, the work in [9] estimates that 80% of disaster victims have their mobile phones present with them at the time of rescue.

The technology discussed within this area of research is quite varied, and it is difficult to categorize implementations into well-defined sub-categories. One interesting observation from the radio-based technologies presented in [10], is that the implementations can be categorized in a fashion analogous to that of the layered OSI model. There are solutions which solely utilizes the physical paradigm, whereas some implementations utilize higher-layer paradigms analogous to that of the data-link and network layers of the OSI model. An example of an implementation which works solely in the physical paradigm is one which reads RSSI of a certain frequency but does not interpret the signal to analyze higher-layer metrics in the signal protocol (such as RSRP).

While cell phones can be located by the cellular provider using technologies available at the base station (i.e., triangulation), this is not guaranteed to work in inhomogeneous environments (i.e., earthquake rubble). Additionally, network carriers are generally reluctant to provide customer location information due to privacy concerns and confidentiality [11]. In the case of post-disaster localization, network-assisted localization is not really an option.

Whilst most literature covered in this chapter utilizes cellular technologies for search and rescue, many of these works employ cellular technologies alongside other technologies to enhance performance.

One such example is utilizing UAVs alongside cellular technology to enhance search and rescue performance. From researching the literature in this field, there is a large emphasis on using UAVs alongside cellular implementations in order to maximize localization performance. It is clear that using mobile platforms (i.e., moving cellular base stations) provides greater operating range and performance. As aerial-based vehicles are the least-impeded form of vehicles, it is obvious that UAVs are being used for such systems. In addition to smart phones and radio technologies, the work in [8] shows that UAVs are a great candidate technology for integrated Urban Search and Rescue systems. It is therefore established that the two main axes that this literature review will focus on are radio technology (cellular in particular) and vehicle technology (UAVs in particular).

Considering cellular radio technology and UAV technology as the two primary axes for this literature review, we can split the literature into four subclasses:

1. Implementations which utilize cellular radio technology *and* UAV technology
2. Implementations which utilize **non-cellular** radio technology *and* UAV technology
3. Implementations which utilize cellular radio technology and **non-UAV** technology
4. Implementations which utilize **non-cellular** radio technology and **non-UAV** technology

The rest of this chapter is organized as follows. Section 2 reviews radio considerations for using wireless transmissions for Search and Rescue purposes. We look at the architecture of a network used in Search and Rescue, the concerned radio metrics used for localization and their processing to extract meaningful information. After establishing these important principles, we explore their use from an Unmanned Aerial Vehicle (UAV) perspective in Section 3. Finally, in Section 4, we explore similar technologies for Search and Rescue that can work when a UAV is not readily available.

2. Radio Considerations for Search and Rescue

2.1 Fundamental Metrics

Literature in this area tends to use a mix of different metrics from which to infer the location of RF targets. The two most prevalent axes for localization metrics are signal time and signal strength. Time based metrics can be broken down into many subcategories (i.e., Time-Difference of Arrival, Time of Flight, Time of Arrival), as can signal-strength metrics (i.e., Received Signal Strength Indicator, Signal to Noise Ratio, Reference Signal Received Power). There are many respective advantages and disadvantages to time and signal-strength metrics for localization. Some common metrics used in such implementations include RSSI and TDoA.

By definition, Received Signal Strength Indicator (RSSI) is the power of a transmitted signal as measured at the intended receiver. This is usually a function of the transmission power and the path loss between the transmitter and receiver. Time Difference of Arrival (TDoA) is the difference in arrival time of a transmitted signal as measured by multiple synchronized receivers at known locations.
Whilst time-oriented metrics, such as Time of Arrival, appear to provide the most accurate location results [12], there is significant computational complexity involved in processing time-oriented metrics for location inferences. Although signal-strength based metrics are inherently less computationally expensive for localization inferences, significant post-processing can be done with these in order to extract location information (i.e., characterizing propagation models). The methods presented in [13] use particle filters for both RSSI (Received Signal Strength Indicator) and AoA (Angle of Arrival) from an aerial point of view to localize RF sources. Another paper focusing on RSSI based localization uses a Kalman Filter to reduce the localization error [14]. Papers like this provide promising improvements to RSSI based localization, which is notoriously coarse. The authors of [15] depict a system in which RSSI and AoA are combined in an aerial sensing fashion with custom path planning and algorithms to find Wi-Fi access points (the targets).

One area of research which bears similarity to post-disaster victim localization is localization of workers in underground mines. The similarity mainly lies in the similar radio propagation environments.

Localization of workers in underground mines is an important aspect of health and safety. Both [16] and [17] are recent papers concerned with RSSI-based localization of mine workers. [17] expresses the prevalence of RSSI based techniques in inhomogeneous underground environments where significant signal propagation challenges are present. The paper also references many papers which attempt to use time based techniques. [18] discusses the inapplicability of time-based metrics to inhomogeneous mine environments, due to the inherent signal physics issues that are present in such environments. It is not a far stretch of the imagination to see that inhomogeneous mine environments bear a lot of similarity with post-disaster rubble environments. Within RSSI based mine personnel localization, there is a range of solutions proposed. [19] proposed a weighted centroid-based algorithm based on a WSN (Wireless Sensor Network) within the mine network, obtaining favorable results. However, [18] comments that there is a significant instability with the system proposed in [19]. [18] goes on to explain the superiority of RSSI based methods for localization in mine environments. The author explains that proposed methods suffer from inherent environmental factors (such as multi-path, diffraction, and shadowing from obstacles). There are many methods to account for these fluctuations and inconsistencies, the most accurate methods being Gaussian location algorithms [18, 20].

2.2 Processing Radio Metrics

Regardless of the radio technology used in the search and rescue system, there will be some processing of radio metrics in order to inform the location of target radio devices. These metrics vary depending on the implementation and can be based on different characteristics of signals, such as frequency, time, and strength.

The processing of these metrics will differ depending on the type of metric. Signal-strength measurements are often used due to their intuitive variation with respect to transmitter and receiver distance (in free space conditions). However, signal-strength measurements can be easily confounded by signal physics issues, such as multipath and shadowing.

For the sake of simplicity, this literature review will focus primarily on signal-strength based radio metric processing. Due to the coarse nature of RSSI, it is often difficult to characterize position as a function of RSSI due to the seemingly non-deterministic nature of RSSI measurements in inhomogeneous environments. However as [21] mentions, RSSI generally scales better as a localization metric than time-based methods in such environments. This is because of the relatively simple hardware requirement of RSSI measurement as compared to the complex hardware of time-based measurement.

RSSI measurements often need to be pre-processed before being fed into positioning algorithms. This is to remove random fluctuations which are often present in continuous RSSI measurements. Oftentimes, a channel model is used to characterize distance as a function of RSSI. If there are multiple receivers measuring the RSSI, then these extrapolated distances can be used to trilaterally localize the RF target. The filtering step must occur before the readings are fed into the localization algorithm. For a filtering algorithm, there are a few important qualities that must be upheld in order to be efficient, such as small computational overhead, and minimal delay. This is to ensure the real-time nature of the localization process. Several filtering algorithms are available, for example (adapted from [21]):

- Moving average filter: Fixed length with an equal weight assigned to each sample. This filter is suitable in situations where individual samples are *not* relatively important.
- Exponential moving average: Fixed length with exponentially decreasing weights applied to old samples. Useful for fixed nodes which 'self-localize'.
- Moving median: Similar to moving average, except uses median. Suitable for situations in which there are many erroneous readings (e.g., heavy NLoS).
- Moving mode: Similar to moving average and moving median. Better suited to unimodal RSSI environments.

These algorithms are compared to the Cramer-Rao Lower Bound. The Cramer-Rao Lower Bound gives the lower bound for the variance of an unknown, deterministic parameter [21, 22]. In this context, the parameter is distance between a transmitter and receiver calculated from RSSI. The authors of [21] express that the simple moving average and exponential moving average filters suffer less from RSSI fluctuations than the moving median and moving mode filters. It is also found that the moving mode filter has the least favorable performance for range estimation. The authors also express that the moving average based filters have better computational efficiency, as they can be computed in real time, O(1), in contrast to moving mode and moving median, which are O(n) and O(n log n) respectively [21]. It was found that the equipment used as well as the environment (Line of Sight or Non Line of Sight) directly affects the RSSI distribution which in turn affects the filtering performance. Nevertheless, filtering appeared to improve the range estimation over using raw RSSI measurements.

2.3 Network Infrastructure Support

4G/LTE is widely regarded as the most prevalent cellular implementation at the time of writing. The 4G/LTE architecture consists of two main components: Evolved Packet Core (EPC) and Radio access network (RAN). The latter is commonly abstracted as the evolved Node B (eNB).

In most implementations, the radio access network is connected via a high-capacity wired link to the EPC. This connection is critical because the EPC performs most of the packet routing and 'heavy-lifting' in the network. The EPC is also the interface between the cellular network and the internet. In some cases, the EPC is connected to the eNB via a high bandwidth microwave link. Yet the consistent theme across implementations is a high bandwidth link between EPC and RAN. However, given the constraints of UAVs, which include limited battery life, and limited payload, it is important to minimize the weight and power draw of the onboard equipment. Therefore, there is an ever-increasing need for more powerful and power-efficient single board computers which can satisfy these constraints and provide enough computing performance to satisfy the cellular network requirements.

One solution to this problem is to separate the eNB and EPC, and have the EPC running on a ground-based computer. This is discussed extensively in [23], whereby the authors propose a UAV based eNB and a ground-based EPC. The link between these two systems would be a steered microwave antenna. This allows the directional microwave antenna to 'follow' the UAV as it flies. The authors of [23] propose this implementation for enhanced LTE connectivity as opposed to a dedicated SAR system. In the use-case of increased LTE connectivity, the backhaul link is even more critical due to the increased traffic that will flow over the said link.

2.4 Other radio considerations

Systems which use RSSI as their method of localization often have the issue of characterizing distance as a function of RSSI, especially in inhomogeneous environments. This could be due to phenomena such as log-normal shadowing due to obstacles in the propagation environment, as well as multipath effects [24]. According to [24] and [25], high bandwidth Time-of-Arrival localization is an attractive metric because of the fine time resolution provided by high-bandwidth ranging signals. This is due to the Fourier relationship between frequency and time. The limitation of high bandwidth Time-of-Arrival localization is apparent in its namesake, high bandwidth. High bandwidth is a rarity considering the congested nature of RF spectrum. Additionally, it is advantageous to utilize technologies that already exist on the user's device, such as Wi-Fi and cellular technologies.

A common method of estimating the direction or angle of arrival of a received signal is through a phased array. As mentioned by [26], phased array systems require very complex signal processing algorithms, such that the receiver is a coherent receiver. Processing of data from phased arrays is known to be an inherently computationally expensive process. Additionally, phased arrays require the use of multiple antennas. It is also important to note that antenna size generally gets larger with lower frequencies, which poses issues at lower cellular frequencies, such as 700 MHz (4G), where antenna size can be prohibitively large.

Antenna design is recognized as a fundamental part of signal-strength based geolocation. Many papers settle on using omnidirectional antennas and relying on post-processing for resolving any ambiguities that arise. However, well-designed directional antennas have the advantage of removing much of the post-processing required. Another research area that bears similarities to this research area, is localization of RF collars worn by endangered wildlife. Many parallels can be drawn between the two research areas. Of particular note, is the shared goal of locating an RF source accurately. One such example of wildlife localization using RF collars is [27] who expresses the applicability of UAVs to RF wildlife tracking. This is especially apparent considering the traditional method of traversing outback terrain on foot using a hand-held antenna and spectrum analyzer. The authors also used Ansys Electronics Desktop to design a custom antenna for aerial wildlife tracking. The paper goes on to discusses the advantages of equipping their SDR system with a bandpass filter to suppress unwanted noise, however they rationalize the lack of filter because of the geographical remoteness of their experiments. The authors state that a bandpass filter would be required for an urban deployment. The authors also state the importance of using other beneficial radio components such as a Low Noise Amplifier (LNA) for improving the Signal to Noise Ratio of the target frequency with respect to the noise floor. This is important in order to create discernible radio hotspots from the noise floor.

Additionally, the propagation environment for the victims' mobile phone is equally as impactful. In the context of earthquakes, there have been a number of studies which attempt to characterize radio propagation in different buildings and building materials prior to, and after collapse [28-35]. With a transmitter placed based inside the buildings and a receiver placed outside the buildings, it was found that signal attenuation fell in the range of 10-30dB for a range of test frequencies. These test frequencies included 900 MHz and 1800 MHz bands, which are common cellular bands. However, after the collapse, this attenuation was as high as 70dB. This speaks to the inhomogeneity of radio signals in building rubble. Other confounding effects for signal propagation include multipath and victim orientation. Multipath occurs alongside propagation attenuation and is incredibly deceiving when trying to locate the source of a transmitted signal. The orientation and location of the victims' body relative to the target mobile phone is also an important factor. For a range of different test frequencies, it was found that there was an average signal attenuation of 20 dB across a range of test subjects when the path between transmitter and receiver is obstructed by the body [36]. It is worth noting that the test frequencies also include 900 MHz and 1800 MHz, which are typical of commercial cellular frequencies.

3. UAV based Search and Rescue Systems

It is undeniable that the proliferation of UAVs has afforded a whole range of services that were previously too logistically difficult to implement. Not so long ago, the idea of attaching a cellular base-station to a UAV was considered a pipe dream, however there is now a whole area of literature concerned with the attachment of cellular femtocells to UAV's. These papers investigate the logistics, performance, and capacity of aerial-based femtocells. These papers do not focus on localization via aerial-based femtocells, this is a whole other research area.

Aerial base stations have been a prominent element in telecommunications research within the last 10 years. This is due to the demand of such systems due to:

- Network rehabilitation after natural and man-made disasters [37-41]
- Temporary cellular capacity increase in densely populated areas
- Enhanced network performance for fast-moving users [42]

For the reasons listed above, aerial base stations continue to be extensively studied. However, there is limited research towards the application of aerial base stations to post-disaster localization, or more generally, search and rescue.

Yet there are other UAV based systems which attempt to sense RF targets based on other radio technologies such as Wi-Fi. There are also papers which rely on raw radio measurements, such as signal strength measurements of a certain frequency from a UAV-based receiver. This variation in implementations reinforces the trend of papers analogous to the OSI model.

In UAV systems, an optimal trade-off has to be found "between flight performance, sensors, and computing resources" according to [43]. The work in [44] emphasizes that land-based robots equipped with wireless sensors have many issues with traversing terrain and obstacles, which can be mostly ignored with an aerial solution. In reinforcement of this narrative, [44] notes that one of the primary advantages of UAVs is the unparalleled mobility in being able to avoid treacherous terrain, whilst also being able to get close to precarious structures with little risk to rescuers. The work in [44] also notes that there is a distinct lack of research (in disaster operations management research) which addresses search for injured people. This is further reinforced by [45]. It is agreed by [44] that UAVs should be able to fill this gap in research. Research reported in [46] addresses emergency management procedures and recognizes that the first 72 hours following a disaster are the most important for the rescue of survivors. Therefore, speed, mobility, detail, and efficacy are hallmarks of good emergency victim localization system. The work in [46] also recognizes that aerial surveying provides the most effective situational awareness.

3.1 UAV SAR systems using cellular radio technologies

In a holistic sense, a UAV-based cellular base station has the advantage of being mobile, and moving towards the UE's which may be buried, allowing for a connection to be established and maintained. Unlike static base stations which may be rendered useless due to the severe attenuation properties of such scenarios [47] or physical/structural damage.

As mentioned before, cellular network deployment via UAVs is a prevalent research area, however, the research area concerned with localization of mobile phones via aerial base stations is gaining traction. One of the first papers to implement such a prototype is detailed in [48], in which a BladeRF x40 Software Defined Radio is paired with a cellular stack (YateBTS) to implement a flying GSM network. This SDR is mounted to a drone for the purpose of finding victims after calamitous events. This system localizes sources using a Levenberg-Marquardt localization algorithm. The depicted system also uses a pair of omnidirectional antennas. While this paper does not discuss any results, it focuses more on the implementation of the system. An interesting observation from this paper (and its successor [49]) is that the authors purposefully augment the SDR frontend such that it appears more powerful to GSM phones in the search area, such that target phones prioritize the powerful GSM station. Another interesting observation is that association requests sent to the UAV-mounted GSM cell from the target cell phones are not accepted so that the cell phones can maintain their connection to the original network. The connection that is established with the airborne GSM base station is adequate to measure RSSI. RSSI is the metric that is fed into the aforementioned localization algorithm. The field-testing results discussed in the successive paper [49], show that the system can localize targets to within a 1m range.

Another paper investigates the attachment of a 4G device to a drone, instead for the purpose of finding victims after avalanches [50]. This paper discusses the unique advantages of UMTS and LTE standards over previous mobile generations for localization purposes. The primary advantages are that LTE deployments employ separate technology standards for the uplink and downlink, therefore making these signals more discernible. Additionally, 4G/LTE takes advantage of better beam-forming technology which allows for inherently better localization due to the more focused downlink signal [50]. This paper also depicts a unique implementation consisting of a custom developed Android app to automate the transmission of LTE parameters from the buried device. This app checks for changes in RSSI (Received Signal Strength Indicator), RSRP (Reference Signal Received Power), and SNR (Signal to Noise Ratio) on the buried phone and sends these changes via email to the 4G device attached to the drone. The 4G device attached to the drone has a camera, from which the operators could see from the point-of-view of the drone. One of the fundamental requirements of this study was to have two devices, one buried and one not buried and use the results from each to inform propagation models and resulting location inferences. One of the primary results of this study is a propagation model for both snow and air. It was also found that propagation of 4G LTE signals through snow was quite good. The system depicted in this paper was interesting due to its application, and the resulting hybrid propagation model.

Another paper which employs a similar system is presented in [51], in which a 4G femtocell is attached to a drone for the purpose of finding disaster victims. This paper appears to be an advancement of [47] and [52]. The method presented in [47] is based on two parameters which are reported from the UE to the eNB in LTE networks. These parameters are RSRP (Reference Signal Received Power) and RSRQ (Reference Signal Received Quality). These measurements are performed for serving cells and neighboring cells when the UE is moving between cells. In commercial LTE deployments, RSRP and RSRQ are usually reported to the eNB from the UE in the case of an LTE A3 event [53]. An LTE A3 event usually occurs when the UE detects a higher received power from a neighboring cell. Unlike the system presented in [54], the system presented in [47] does not require a cellular jammer for existing infrastructure. The UE is connected solely to the aerial femtocell. In [47], there are two main phases of operation, the classification phase and localization phase. The classification phase attempts to determine whether a UE is inside the specified localization area. The localization phase attempts to find the location of the UE. Once the mobile phone is connected to the femtocell and the localization phase has started, the RSRP is read from the femtocell, and mapped according to its location, resulting in a heat map of RSRP. Further post-processing algorithms, such as weighted distance method, center of gravity method, and other methods are executed dependent on the characteristics of the heat map, namely the number of "lakes" and "mountains" (low and high RSRP respectively) within the localization area. The UAV is flown half a meter above the maximum rubble height. Interestingly, the researchers found that a lower power was observed when the drone was directly above the UE as opposed to areas slightly offset from the UE. This is due to the vertical polarization of the omnidirectional femtocell antennas used [47], whereby the radiation pattern of the antenna does not correlate well with the uplink signal polarization. The testing carried out in [47, 52] do not use a drone, instead they move the femtocell manually and scan walls of a building as opposed to a rubble site. However, the latest paper published by the authors [51] tests the proposed system on a UAV platform. The results from this study show that a localization accuracy of just above 60% with an average positioning error of 1m. The authors of [55], claim to the best of their knowledge, that their detailed system is the first 'drone-based cellular search-and-rescue solution'. If the publication date is anything to go by, this claim is false. The solution, dubbed 'SARDO' uses technologies such as time-of-flight, machine-learning, feedback-based trajectory management and IMSI (International Mobile Subscriber Identity) interception. The fundamental principle of operation of SARDO is as follows: Time of Flight measurements are performed between the cellular-enabled SDR and the UE, these measurements are fed into a convolutional neural network which extrapolates relevant details from Time-of-Flight measurements to feed into a deep Feed-Forward Neural Network which "learns and implements the concept of pseudo-trilateration". The combination of this CNN and the subsequent

LSTM (Long Short-Term Memory) network is used to gauge motion trajectory for future position detection. SARDO is only designed to localize UE's sequentially, not simultaneously. The concept of pseudo-trilateration is interesting and is relatively easily afforded in UAV systems. The idea of pseudo-trilateration is to travel between multiple pre-defined locations within a specific window of time to accumulate the data used for subsequent trilateration. This seems like an efficient alternative to having three or more time-synchronized UAV systems. [55] operates by deploying an LTE cell with omnidirectional antennas. The results of this paper claim that SARDO can determine the location of mobile phones at a rate of 3 minutes per UE and achieve an accuracy of tens of meters.

From the aforementioned selection of papers, it can be seen that there is a rising trend in cellular-based UAV SAR systems.

3.2 UAV SAR systems using non-cellular radio technologies

While there is a whole research field concerned with aerial base stations, there are other radio technologies that can be deployed on a UAV platform for search and rescue purposes. One prevalent implementation amongst this subset of research is localization of Wi-Fi enabled devices through Wi-Fi probe requests [56-58]. The results from one such study [56] show that devices can be reliably detected from up to 200m aerially. One disadvantage of this particular system was that, for best performance, the MAC address of the Wi-Fi device needed to be known prior to flying. The intended use case for this system is wilderness scenarios where the propagation environment is relatively homogeneous.

There are also other aerial systems which attempt to localize RF targets via Wi-Fi probe requests. Another example is [57], which sniffs for probe requests from mobile phones. The RSSI from these probe requests is extracted and processed through a Kalman filter to remove fluctuations due to multi-path and other signal physics issues. This RSSI is used to calculate distance based on propagation models. It then finds a direction 'gradient' using a circular flight path. A localization accuracy of 7m was achieved in this study.

A similar technique is shown in [59], except the post-processing workflow consists of machine learning networks. There is also more geographical pre-processing involved in this paper, resulting in overall estimation accuracy of 81.8%. An interesting point mentioned in this paper regards the flight path priority. The authors mention that the UAV should fly sufficiently slow in order to maximize the effective sampling rate. However, due to the time-sensitive nature of the operation, the UAV should also cover this area as quickly as possible. This is the trade-off between flight area and quality/accuracy that is evident across most research in this field.

The system depicted in [60] attempts to localize RF devices irrespective of their operating protocol, by using antenna switching. The system uses four separate low-gain directional antennae to localize a transmitting RF source based on intermittent bearing estimation. Whereby the drone will find the bearing of the RF source, then fly towards the source for a fixed distance. This algorithm repeats until the transmitting source is found. The authors of [60] extended the scope of the project to find the optimal location between two ground users for investigative purposes. It is worth noting that the authors only attempt to target a transmitting SDR and not a mobile phone. This is a limitation of the study and is an opportunity for further exploration. The exploitation of antenna switching between the four directional patch antennae proves to be a good method for estimating the transmission source bearing in relatively homogenous propagation environments.

The main contribution in [61] is demonstrating a two-antenna system for pseudo-bearing measurements, this paper focuses on a Wi-Fi implementation. The system uses a 9 dBi Yagi-Uda directional antenna for bearing measurement and an omnidirectional antenna for normalizing bearing measurements. The paper validates the normalization theory, and the paper concludes that the proposed system is faster at locating RF sources than other systems which exploit 'rotate for bearing' modalities. Another paper [62] uses low-gain directional antennae (Moxon antennae) paired to an RTL-SDR radio to localize a variety of sources, including a handheld radio, an RF-enabled collar, and a smartphone. In this paper, the smartphone is forced into a phone call to ensure that it is constantly transmitting a signal. This method seems somewhat unrealistic for a disaster scenario. However, this brings up an important point regarding 'inducing' cellular signals in order to localize them. There needs to be an active or intermittent uplink signal from the mobile phone in order to reliably detect the location of the mobile phone. Examples of scenarios where an active uplink signal is present include a phone-call or video streaming. An ideal solution would remove the need for the user to initiate such a transmission. Therefore, mechanisms for inducing uplink transmissions from a target mobile phone needs to be established. A good example of such a mechanism is Internet Control Message Protocol (ICMP), which is more commonly known as 'ping'. Pinging the IP address of the target mobile phone would induce an uplink signal from the mobile phone. Perhaps implementing a cellular stack (paired with an SDR) would allow for a more integrated system and potentially better localization results in this study. However, the main contributions of this paper are to create a general RF localization tool that is faster than existing systems. [63] exploits the use of directional antennae for the localization and tracking of RF sources by measuring RSS (Received Signal Strength). This system uses a particle filter approach to inform the path planning algorithm. The directional antenna used is a PCB based quad-patch antenna. The system is sensing using a pair of 2.4GHz XBee transceivers.

One paper has presented a novel algorithm which leverages an array of directional antenna from a fixed wing UAV platform to localize targets in sparse sensor networks [64]. The authors claims that the use of directional antenna means that each sensor only requires one trilateration for localization, therefore optimizing the workflow over omnidirectional implementations. Additionally, by using directional antennas, rough estimation of the sensors position does not need to be performed initially. This contrasts with omnidirectional implementations as well.

It can be seen from the aforementioned literature that there are a few key variable components in each implementation, such as RF target type, antennas used, and specific deployment environments.

3.3 UAV Path-planning algorithms for SAR

Many UAV based systems also present path planning algorithms only. This is a large aspect of any UAV based localization system. The intuitive approach would be to use war driving (or more appropriately, war flying). However, there are many different approaches to this.

The system proposed in [65] uses D2D (Device-to-Device) communication and 3GPP proximity-based services (ProSe) for estimating the number of victims in the search area. The paper justifies development based on the claim that existing PSC (Public Safety Communication) networks are not suitable for emergency responses. D2D and ProSe complement each other in disaster scenarios, ProSe allowing discovery and communication between mobile devices, and D2D providing a realizable full coverage solution for mobile devices. The authors in [65] also propose an implementation of UAV mounted LTE base station. The authors of [65] also claim that UAV mounted base stations have the fundamental limitation of core network connection. This builds upon the UAV backhaul issue mentioned previously. Unlike other papers, the work in [65] purports to "accelerate the rescuing process in emergency responses". This contrasts with precise localization of devices. The paper proposes a four-drone implementation, by initially narrowing the search window with the drones, the number of devices in each 'emergency area' are counted based on the devices with D2D connections. The paper goes on to analyze and attempt to correct errors in victim counting algorithms. This paper seems to be based purely on simulation with no practical implementation.

The work in [66] proposes an algorithm for localizing RF targets based on RSSI sensing in real environments. The authors claim that existing research which relies on localization via RSSI alone does not account well enough for situations in which the target may be covered by obstacles. The authors in [66] propose to use the deviation of RSSI as the UAV traverses the terrain. A visual depiction of the algorithm is shown in Figure 1. The paper focuses on the implementation and analysis of the algorithm. From the scanning area, a smaller 'scanning' square (denoted by S_n) is set, with one of those vertices being the starting RSSI measurement point. The length of one side of the square is denoted as d. The UAV traverses the vertices of the square, measuring RSSI at each vertex. The vertex with the maximum RSSI is defined as the symmetric point, from which a new square is set. This new square is set such that the new square and the previous square share the vertex with maximum RSSI. The deviations between square S_n and S_{n-1} are then calculated. The distance d is changed if there is a large difference. Once the measured RSSI reaches a threshold (-50 dBm in this case), the location of the target is said to be found. There is a tradeoff between search time and success rate, depending on the value of d. The authors state that, with a value of d=2.5m, the success rate was 90%, however, the search time was high. It was found that a value of d=10m resulted in the lowest search time.

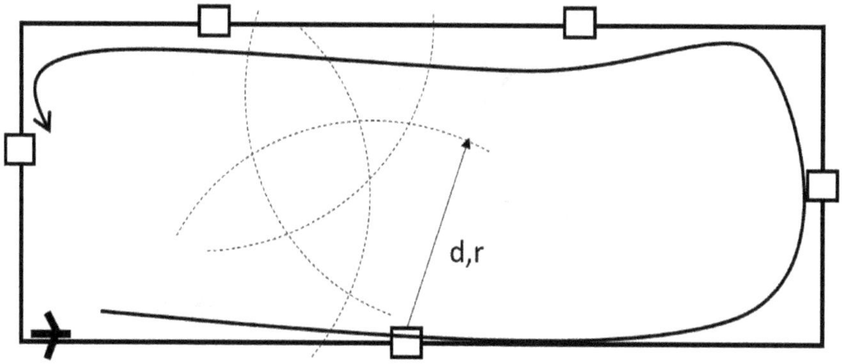

Figure 1: The flying algorithm proposed in and abstracted from [66].

Whilst not based on *conventional* disaster sites, [67] focuses on maritime search and rescue through UAV based localization. The proposed system equips a UAV with the capability to listen to mobile communications when the mobile device is still connected to mainland cellular base stations. The UAV also has the capability to deploy a cellular network to communicate with mobile targets. The paper proposes that the base station and the mobile target will communicate using silent SMS in order to get regular RSSI updates. This can then be used to inform the localization algorithm. This paper focuses on an NLLS (non-linear least squares) approach to localize the mobile target from the virtual anchor nodes on the predetermined UAV flight path, by measuring RSSI at these nodes. A visual depiction of this is shown in Figure 2.

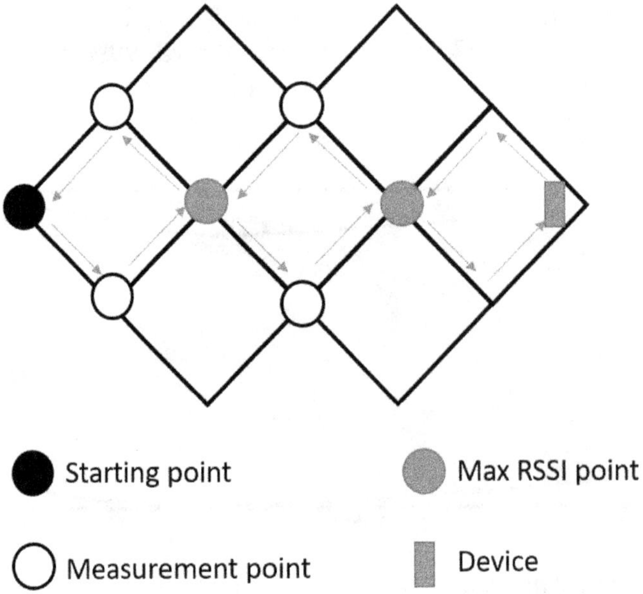

Figure 2: The algorithmic implementation proposed in and abstracted from [67].

Along with the parallel sweep path planning method (as seen in Figure 2), the paper also investigates other path planning methods such as expanding square, which starts from a central location and expands outwards in a rectangular fashion. Lastly, the sector search is investigated, whereby the UAV flies outwards (from the center of the search area) at certain angles in order to localize mobile targets within the defined sector. There is more to the algorithms mentioned in this paper. However, these other algorithms are outside of the scope of this literature review. Efficient path planning algorithms are an incredibly important aspect of automated UAV operations, especially with regard to battery life.

Another flight path algorithm which appears to be standard in capture of images for photogrammetry and creation of orthomosaic is *grid-style* flight paths. In flight planning software, this style of mapping requires the user to input the desired height at which to map the selected area, and also the longitudinal and latitudinal overlaps for sufficient orthomosaic stitching performance [68]. For search and rescue, grid-style mapping provides excellent coverage [69], at the cost of very large datasets. Longitudinal and latitudinal overlaps of 80% are recommended by [69] in order to capture sufficient detail for visual search and rescue.

3.4 Special Considerations for UAV implementations

While there are a number of UAV-based solutions mentioned above, there are a few limitations of UAVs at the time of writing. Whilst most UAVs have powerful flight computers, they do not necessarily support conventional computation and GPIO. Therefore, in order to capture data and perform processing with third-party sensors (i.e., SDR) onboard the drone, an embedded computer will have to be mounted to the drone. Unfortunately, all of the added componentry decreases the battery life of the UAV, due to power consumption, and added weight. Therefore, an optimization problem is born. Whereby the solution must strike a compromise between flight-time and onboard features. An ideal solution would mount the bare minimum componentry aboard the UAV while moving non-essential processing to a ground station.

Of course, some of the inherent issues with UAVs in this context is battery life and the ability to maintain a backhaul link (with existing infrastructure). There are a number of papers which propose solutions to these issues [70-72]. [72] mentions the advantages of 'tethering' UAVs to stable power sources and/or high-speed backhaul data links. This allows for seemingly unlimited flight time and unimpeded network performance. However, the mobility of such a UAV would be limited by the tether. [72] proposes a grid array of non-tethered UAVs for the purpose of network rehabilitation after large-scale natural disasters. In addition to the grid of UAVs, [72] proposes three types of drones, a tethered backhaul drone, a powering drone, and a communication drone BS. In the implementation, the communication drones construct the grid, while the tethered backhaul drone communicates with the grid of communication drones via RF/FSO link, and the powered drone visits each drone to charge it when required. This is a multi-drone implementation which takes advantage of the mobility of untethered drones, yet also harnesses the advantages of tethered drones. While [72] only considers the mathematical optimization of such a network, the implementation would be very difficult. Therefore, for the purposes of localization in rough post-disaster scenarios, it is important to explore the most flexible mobile solutions (i.e., untethered).

In a similar fashion to vehicle-based networks such as cell-on-wheels (COW), aerial based networks require sufficient backhauling in order to make the system effective as a communications relay. In the context of localization, however, there is not much need for data backhaul. This is because the primary purpose of the aerial base station is to locate the target UEs, not enable large data transfers. A non-backhauled system is known as a standalone system. However, due to the computational complexity of running a full cellular stack (RAN + EPC), there is a large power draw, rendering such a system non-optimal for an untethered UAV. Therefore, if some of the processing can be relieved from the UAV, and shifted onto a more capable ground station computer, the drone will have more resources to remain in flight. Therefore, instead of backhauling the internet connection, the constituent components of the cellular network can be separated and backhauled (i.e., EPC - RAN backhaul link). Both [71] and [70] propose antenna steering backhaul links between ground stations and drones for the purpose of using the drone as a communications relay. [23] introduces the challenges of designing a flexible, untethered aerial base station. As noted by [23], the rate at which the radio channel characteristics change for the RAN-UE connection is much higher than the EPC-RAN backhaul link. This is an important consideration to take into account when considering the optimization of backhaul and drone placement. The communication mode of the EPC-RAN backhaul link is also an important consideration for optimization. As mentioned in [23], antenna size can be prohibitive for such a backhaul link, especially at lower frequencies. At high frequencies, the 'directionality' of backhaul antenna needs to be much higher due to the greater propagation losses. The clear choice appears to be Wi-Fi, this is due to the widespread compatibility of Wi-Fi with most SBC computers. However, Wi-Fi operates in unlicensed spectrum, which can be congested in busy areas. The intricate details of different backhaul communication modes are further discussed in [23].

Another issue which impacts backhaul quality is the attitude orientation of the UAV. This is because the orientation of the backhaul antenna also changes. The effects of this can be somewhat mitigated by using an omnidirectional antenna on the UAV. This is because the polarization of an omnidirectional antenna would not change drastically during different movement modalities. However, having a non-directional backhaul antenna on the UAV impinges on the link quality.

Whilst a transmitted signal from an overhead aerial source would be inherently less susceptible to shadowing and path-loss from the perspective of a ground-based receiver, there are other considerations, such as the UAVs control frequency, noise from motors/ESCs, and so forth. [27] also mentions the importance of investigating the noise that may be present from the UAV itself (i.e., the ESCs and motors). By measuring the noise floor from an RTL-SDR with the drone operating, the authors concluded that it was inconclusive as to whether the drone was contributing to the measured noise floor [27].

Additionally, systems such as aerial base stations must ensure that there are sufficient uplink and downlink radio propagation characteristics in order to ensure good communication. One such method to ensure this is to use directional antenna aboard the aerial base station. Using directional antenna on the aerial base station can significantly increase the transmit and receive gain in certain directions. Additional components such as power amplifiers, and low noise amplifiers can also assist in increasing transmit and receive gain. Most directional antennas are provided with a datasheet of their RF characteristics, such as gain, beamwidth, and radiation pattern. These are all useful parameters for planning a network. In order to properly plan the coverage of an aerial base station, it is important to consider how the beamwidths of the antenna determine the placement of the drone, in terms of height and position. This consideration is reinforced by [23]. Looking further into this problem, as the drone moves (i.e., pitch, yaw, and roll), the coverage pattern on the ground will change if the directional antenna is mounted statically aboard the drone. Some of these effects can be mitigated if the directional antenna is gimbaled and forced to remain nadir (i.e., always pointing at the ground directly below the drone, regardless of pitch, yaw, and roll). Nadir is a familiar term in UAV camera-based mapping, whereby the gimbaled camera performs compensation base on the pitch, yaw and roll of the drone. This will allow the coverage pattern to move linearly with the spatial position of the drone. There appears to be a lack of implementations which deploy directional antennas aboard UAVs, let alone nadir antennas.

4. Non-UAV based Search and Rescue systems

As mentioned previously, most of the papers in this literature review focus on UAVs due to their fundamental advantages in mobility and perspective. However, it is interesting to explore other systems which utilize cellular radio technologies.

4.1 SAR systems using cellular radio technologies

One example of a SAR system using cellular technology is [73] which uses a hybrid AoA/RSSI (Angle of Arrival/Received Signal Strength Indicator) positioning scheme to localize cell phones. The mechanism by which this is achieved is a ground based OpenBTS-equipped system. Similar to YateBTS, OpenBTS is another cellular stack which can be deployed in conjunction with Software Defined Radios. The proposed positioning process is as follows: The UE's position is coarsely estimated via AoA and a sector antenna. Following this phase, a more directional Yagi-Uda antenna is used for accurate positioning. Some interesting criteria were imposed upon the solution by the authors to make the proposed system disaster-friendly, such as keeping the weight of the entire system below 30kg and making the system easy to carry. This paper makes reference to [74] who propose a similar solution. Similar to the aforementioned paper, the work in [74] also makes use of a static network. However, instead of using antenna modalities to infer the location of the UE, the work in [74] logs the GPS location of SMS messages sent from each UE to the base station. [73] alludes to the fact that [74] does not make use of power devices such as power amplifiers or low noise amplifiers to enhance the localization process, reinforcing the relatively small detection range. The system proposed in [73], sets up a localized GSM network using a USRP SDR, laptop, LNA, PA and battery. The system has a maximum transmission power of 20W allowing a larger detection range than [74]. The authors use a long open road as a testing site instead of a disaster site. When testing the Yagi for accurate positioning, it was found that three parameters changed with respect to the direction of the antenna. These were RSSI, MSP (Mobile Station Power) and LoS (Line of Sight). Intuitively, it was found that if the antenna faces the direction of the mobile device, the MSP and LoS decrease, and the RSSI increases. A large limitation of this study is the handheld nature of the Yagi antenna. This means that the time taken to locate victims may be quite long due to the arduous nature of traversing post-disaster terrain. There is, perhaps, an opportunity to attach this Yagi antenna to a vehicle for more optimized searching.

There are relatively few non-UAV systems which utilizes cellular technologies for SAR purposes. Another example similar to [73] is [54]. The analysis of this paper was covered more extensively in [7]. However, the premise of the implementation, was that a custom GSM cell was deployed in a portable setup and a GSM jammer was used to jam existing GSM infrastructure. First responders could then scan the area on foot with a handheld 'power meter' and sense areas of high signal strength.

4.2 Resulting novelty

Within the research field of localization of ground based UEs via aerial base station (or SAR more generally), there appears to be a lack of research in which highly directional nadir antennas are used. This is particularly apparent in the context of simultaneous localization and mapping. Simultaneous localization and mapping is a prevalent theme in autonomous vehicular search and rescue [75, 76]. Additionally, there are very few systems which utilizes higher-layer metrics of the LTE stack, whilst maintaining compatibility with COTS UEs. Therefore, the findings from this literature review warrants the development of a UAV-based system which employs a cellular stack for the purposes of disaster victim localization, utilizing nadir directional antenna and higher-layer signal-strength based metrics.

The context and implementation of such a system within post-disaster management and organization also needs to be considered. This alludes to the use case of such a system within the search and rescue process. Such a system can be used as an initial tool that's deployed which provides location inferences. Alternatively, such a tool can be used more for highly accurate location prediction midway through the search and rescue process. Ultimately, this decision needs to be made after assessing the accuracy of such a system and assessing the needs of search and rescue teams. A paper which focused on the use of UAV systems by dedicated SAR groups sent a survey to said SAR groups and obtained responses on ideal operation characteristics of such a system [77, 78]. It was found that respondents were not overly focused on high accuracy of a system and were looking for a system which would provide early insights on victim location.

The authors are working on developing Search and Rescue systems that reply only on mobile phone transmissions to localize victims with the help of a UAV. Prototyping work for such a system has recently been completed [7, 79], which suggests that after incorporating identified modifications, the use of UAV in picking up signals from victims' mobile phones can be a feasible solution for localization in post disaster environments [64].

Bibliography

1. Spence, R.J.S., E. So, and C. Scawthorn, *Human casualties in earthquakes: progress in modelling and mitigation*. Advances in natural and technological hazards research: v. 29. 2011: Springer.
2. Noji, E.K., et al., *The 1988 earthquake in Soviet Armenia: A case study*. Annals of Emergency Medicine, 1990. **19**(8): p. 891-897.
3. Olson, J.A., et al., *Smartphone addiction is increasing across the world: A meta-analysis of 24 countries*. Computers in Human Behavior, 2022. **129**: p. 107138.
4. Bäckström, C.-J. and N. Christoffersson, *Urban Search and Rescue-An evaluation of technical search equipment and methods*. LUTVDG/TVBB--5197--SE, 2006.
5. Wong, J. and C. Robinson, *Urban search and rescue technology needs: identification of needs*. Federal Emergency Management Agency (FEMA) and the National Institute of Justice (NIJ). Document, 2004. **207771**.
6. Sujatmiko, W., et al. *A Review of Radars to Detect Survivors Buried Under Earthquake Rubble*. in *2017 5th International Conference on Instrumentation, Communications, Information Technology, and Biomedical Engineering (ICICI-BME)*. 2017.
7. Murray, T. and S.F. Hasan. *Present State of the Art in Post Disaster Victim Localization*. in *2020 IEEE 5th International Symposium on Telecommunication Technologies (ISTT)*. 2020.

8. Statheropoulos, M., et al., *Factors that affect rescue time in urban search and rescue (USAR) operations.* Natural Hazards: Journal of the International Society for the Prevention and Mitigation of Natural Hazards, 2015. **75**(1): p. 57.
9. Zorn, S., et al., *A smart jamming system for UMTS/WCDMA cellular phone networks for search and rescue applications.* 2012, IEEE. p. 1-3.
10. Hamp, Q., et al., *New technologies for the search of trapped victims.* Ad Hoc Networks, 2014. **13**(Part A): p. 69-82.
11. Peral-Rosado, J.A.d., et al., *Survey of Cellular Mobile Radio Localization Methods: From 1G to 5G.* IEEE Communications Surveys & Tutorials, 2018. **20**(2): p. 1124-1148.
12. Xu, N., et al. *Machine Learning Based Outdoor Localization using the RSSI of Multibeam Antennas.* in *2020 IEEE Workshop on Signal Processing Systems (SiPS)*. 2020.
13. Dehghan, S.M.M., M. Farmani, and H. Moradi. *Aerial localization of an RF source in NLOS condition.* in *2012 IEEE International Conference on Robotics and Biomimetics (ROBIO)*. 2012.
14. George, M. and K. Vadivukkarasi, *Kalman filtering for RSSI based localization system in wireless sensor networks.* 2015. **10**: p. 16429-16440.
15. Wang, A., et al., *GuideLoc: UAV-Assisted Multitarget Localization System for Disaster Rescue.* Mobile Information Systems, 2017. **2017**: p. 1267608.
16. Yi, L., L. Tao, and S. Jun. *RSSI localization method for mine underground based on RSSI hybrid filtering algorithm.* in *2017 IEEE 9th International Conference on Communication Software and Networks (ICCSN)*. 2017.
17. Ming, J., et al. *Study on the personnel localization algorithm of the underground mine based on RSSI technology.* in *2017 IEEE 9th International Conference on Communication Software and Networks (ICCSN)*. 2017.
18. Ge, B., et al., *Improved RSSI Positioning Algorithm for Coal Mine Underground Locomotive.* Journal of Electrical and Computer Engineering, 2015. **2015**: p. 918962.
19. Han, D.-S., et al., *A weighted centroid localization algorithm based on received signal-strength indicator for underground coal mine.* Journal of China Coal Society, 2013. **38**(3): p. 522.

20. Jianwu, Z. and Z. Lu. *Research on distance measurement based on RSSI of ZigBee*. in *2009 ISECS International Colloquium on Computing, Communication, Control, and Management*. 2009.
21. Koledoye, M.A., et al. *A Comparison of RSSI Filtering Techniques for Range-based Localization*. in *2018 IEEE 23rd International Conference on Emerging Technologies and Factory Automation (ETFA)*. 2018.
22. Paúr, M., et al., *Tempering Rayleigh's curse with PSF shaping*. Optica, 2018. **5**(10): p. 1177-1180.
23. Sundaresan, K., et al., *SkyLiTE: End-to-end design of low-altitude UAV networks for providing LTE connectivity*. arXiv preprint arXiv:1802.06042, 2018.
24. Aditya, S., A.F. Molisch, and H.M. Behairy, *A Survey on the Impact of Multipath on Wideband Time-of-Arrival Based Localization*. Proceedings of the IEEE, 2018. **106**(7): p. 1183-1203.
25. Qi, Y., H. Kobayashi, and H. Suda, *Analysis of wireless geolocation in a non-line-of-sight environment*. IEEE Transactions on wireless communications, 2006. **5**(3): p. 672-681.
26. Hood, B.N. and P. Barooah, *Estimating DoA From Radio-Frequency RSSI Measurements Using an Actuated Reflector*. IEEE Sensors Journal, 2011. **11**(2): p. 413-417.
27. Webber, D., et al. *Radio receiver design for unmanned aerial wildlife tracking*. in *2017 International Conference on Computing, Networking and Communications (ICNC)*. 2017. IEEE.
28. Holloway, C.L., *Propagation and detection of radio signals before, during, and after the implosion of a large convention center / Christopher L. Holloway, et al*. NIST technical note: 1542. 2006, [Boulder, CO]: National Institute of Standards and Technology.
29. Holloway, C.L., *Propagation and detection of radio signals before, during, and after the implosion of a thirteen-story apartment building / Christopher L. Holloway*. NIST technical note: 1540. 2005.
30. Holloway, C.L., et al. *Radio propagation measurements during a building collapse: Applications for first responders*. in *Proc. Intl. Symp. Advanced Radio Tech., Boulder, CO*. 2005.
31. Holloway, C.L., et al., *Propagation and detection of radio signals before, during, and after the implosion of a large sports stadium (Veterans' Stadium in Philadelphia)*. Natl. Inst. Stand. Technol. Note, 2005. **1541**.

32. Holloway, C.L., et al., *Propagation measurements before, during, and after the collapse of three large public buildings.* IEEE Antennas and Propagation Magazine, Antennas and Propagation Magazine, IEEE, IEEE Antennas Propag. Mag., 2014. **56**(3): p. 16-36.
33. Holloway, C.L., et al., *Attenuation of radio wave signals coupled into twelve large building structures.* 2008.
34. Remley, K.A., et al., *Radio-Wave Propagation Into Large Building Structures—Part 2: Characterization of Multipath.* IEEE Transactions on Antennas and Propagation, Antennas and Propagation, IEEE Transactions on, IEEE Trans. Antennas Propagat., 2010. **58**(4): p. 1290-1301.
35. Young, W.F., et al., *Radio-Wave Propagation Into Large Building Structures—Part 1: CW Signal Attenuation and Variability.* IEEE Transactions on Antennas and Propagation, Antennas and Propagation, IEEE Transactions on, IEEE Trans. Antennas Propagat., 2010. **58**(4): p. 1279-1289.
36. Jiao, H., et al., *Radio frequency propagation characteristics in disaster scenarios.* 2014, IEEE. p. 818-821.
37. Gomez Chavez, K., et al., *Capacity Evaluation of Aerial LTE Base-Stations for Public Safety Communications.* 2015.
38. Guevara, K., et al. *UAV-based GSM network for public safety communications.* in *SoutheastCon 2015.* 2015.
39. Deruyck, M., et al. *Emergency ad-hoc networks by using drone mounted base stations for a disaster scenario.* in *2016 IEEE 12th International Conference on Wireless and Mobile Computing, Networking and Communications (WiMob).* 2016. IEEE.
40. Deruyck, M., et al., *Designing UAV-aided emergency networks for large-scale disaster scenarios.* EURASIP Journal on Wireless Communications and Networking, 2018. **2018**(1): p. 79.
41. Radišić, T., M. Muštra, and P. Andraši. *Design of an UAV Equipped With SDR Acting as a GSM Base Station.* in *2019 International Conference on Systems, Signals and Image Processing (IWSSIP).* 2019.
42. Enayati, S., et al., *Moving aerial base station networks: A stochastic geometry analysis and design perspective.* IEEE Transactions on Wireless Communications, 2019. **18**(6): p. 2977-2988.
43. Tomic, T., et al., *Toward a fully autonomous UAV: Research platform for indoor and outdoor urban search and rescue.* IEEE robotics & automation magazine, 2012. **19**(3): p. 46-56.

44. Grogan, S., R. Pellerin, and M. Gamache, *The use of unmanned aerial vehicles and drones in search and rescue operations – a survey.* 2018.
45. Galindo, G. and R. Batta, *Review of recent developments in OR/MS research in disaster operations management.* European Journal of Operational Research, 2013. **230**(2): p. 201-211.
46. Erdelj, M., M. Król, and E. Natalizio, *Wireless sensor networks and multi-UAV systems for natural disaster management.* Computer Networks, 2017. **124**: p. 72-86.
47. Avanzato, R. and F. Beritelli, *An Innovative Technique for Identification of Missing Persons in Natural Disaster Based on Drone-Femtocell Systems.* Sensors (Basel, Switzerland), 2019. **19**(20): p. 4547.
48. Murphy, S.Ó., K.N. Brown, and C.J. Sreenan. *Cellphone Localisation using an Autonomous Unmanned Aerial Vehicle and Software Defined Radio.* in *EWSN.* 2017.
49. Murphy, S.O., C. Sreenan, and K.N. Brown. *Autonomous unmanned aerial vehicle for search and rescue using software defined radio.* in *2019 IEEE 89th vehicular technology conference (VTC2019-Spring).* 2019. IEEE.
50. Wolfe, V., et al. *Detecting and locating cell phone signals from avalanche victims using unmanned aerial vehicles.* in *2015 International Conference on Unmanned Aircraft Systems (ICUAS).* 2015.
51. Avanzato, R. and F. Beritelli, *A Smart UAV-Femtocell Data Sensing System for Post-Earthquake Localization of People.* IEEE Access, 2020. **8**: p. 30262-30270.
52. Avanzato, R., F. Beritelli, and M. Vaccaro. *Identification of Mobile Terminal with Femtocell on Drone for Civil Protection Applications.* in *2019 10th IEEE International Conference on Intelligent Data Acquisition and Advanced Computing Systems: Technology and Applications (IDAACS).* 2019.
53. Mehta, M., N. Akhtar, and A. Karandikar. *Impact of HandOver parameters on mobility performance in LTE HetNets.* in *2015 Twenty First National Conference on Communications (NCC).* 2015.
54. Zorn, S., et al. *A novel technique for mobile phone localization for search and rescue applications.* in *2010 International Conference on Indoor Positioning and Indoor Navigation.* 2010.

55. Albanese, A., V. Sciancalepore, and X. Costa-Pérez, *SARDO: An automated search-and-rescue drone-based solution for victims localization.* arXiv preprint arXiv:2003.05819, 2020.
56. Wang, W., et al. *Feasibility study of mobile phone WiFi detection in aerial search and rescue operations.* in *Proceedings of the 4th Asia-Pacific workshop on systems.* 2013.
57. Sun, Y., et al. *Localization of WiFi Devices Using Unmanned Aerial Vehicles in Search and Rescue.* in *2018 IEEE/CIC International Conference on Communications in China (ICCC Workshops).* 2018.
58. Kashihara, S., et al. *Wi-SF: Aerial Wi-Fi Sensing Function for Enhancing Search and Rescue Operation.* in *2019 IEEE Global Humanitarian Technology Conference (GHTC).* 2019.
59. Acuna, V., et al. *Localization of WiFi Devices Using Probe Requests Captured at Unmanned Aerial Vehicles.* in *2017 IEEE Wireless Communications and Networking Conference (WCNC).* 2017.
60. Petitjean, M., S. Mezhoud, and F. Quitin. *Fast localization of ground-based mobile terminals with a transceiver-equipped UAV.* in *2018 IEEE 29th Annual International Symposium on Personal, Indoor and Mobile Radio Communications (PIMRC).* 2018.
61. Dressel, L. and M.J. Kochenderfer. *Pseudo-bearing Measurements for Improved Localization of Radio Sources with Multirotor UAVs.* in *2018 IEEE International Conference on Robotics and Automation (ICRA).* 2018.
62. Dressel, L.K. and M.J. Kochenderfer. *Efficient and low-cost localization of radio signals with a multirotor UAV.* in *2018 AIAA Guidance, Navigation, and Control Conference.* 2018.
63. Isaacs, J.T., et al. *Quadrotor control for RF source localization and tracking.* in *2014 International Conference on Unmanned Aircraft Systems (ICUAS).* 2014. IEEE.
64. Sorbelli, F.B., et al. *Precise localization in sparse sensor networks using a drone with directional antennas.* in *Proceedings of the 19th International Conference on Distributed Computing and Networking.* 2018.
65. Yu, J. and F. Ye. *User Equipment Localization and Victim Estimation with Next-Generation PSC in Emergency Response.* in *2019 IEEE Global Communications Conference (GLOBECOM).* 2019.

66. Tomiyama, M., Y. Takeda, and T. Koita. *Location estimation algorithm using UAV for real environments.* in *2020 Eighth International Symposium on Computing and Networking Workshops (CANDARW)*. 2020.
67. Tiemann, J., O. Feldmeier, and C. Wietfeld. *Supporting Maritime Search and Rescue Missions Through UAS-Based Wireless Localization.* in *2018 IEEE Globecom Workshops (GC Wkshps)*. 2018.
68. Strecha, C., O. Küng, and P. Fua, *Automatic mapping from ultra-light UAV imagery*. 2012.
69. Weldon, W.T. and J. Hupy, *Investigating methods for integrating unmanned aerial systems in search and rescue operations.* Drones, 2020. **4**(3): p. 38.
70. Pascual Campo, P., *Antenna Steering System For Directional Microwave Link With UAV Communications.* 2018.
71. Pokorny, J., et al., *Concept design and performance evaluation of UAV-based backhaul link with antenna steering.* Journal of Communications and Networks, 2018. **20**(5): p. 473-483.
72. Selim, M.Y. and A.E. Kamal. *Post-disaster 4G/5G network rehabilitation using drones: Solving battery and backhaul issues.* in *2018 IEEE Globecom Workshops (GC Wkshps)*. 2018. IEEE.
73. Tang, S., et al., *Study of portable infrastructure-free cell phone detector for disaster relief.* Natural Hazards, 2017. **86**(1): p. 453-464.
74. Hatorangan, E. and T. Juhana. *Mobile phone location logging into OpenBTS-based cellular network in disaster situation.* in *2014 8th International Conference on Telecommunication Systems Services and Applications (TSSA)*. 2014.
75. Heukels, F., *Simultaneous Localization and Mapping (SLAM): towards an autonomous search and rescue aiding drone.* 2015, University of Twente.
76. Wang, H., et al., *Information-Fusion Methods Based Simultaneous Localization and Mapping for Robot Adapting to Search and Rescue Postdisaster Environments.* Journal of Robotics, 2018. **2018**: p. 4218324.
77. Półka, M., S. Ptak, and Ł. Kuziora, *The Use of UAV's for Search and Rescue Operations.* Procedia Engineering, 2017. **192**: p. 748-752.
78. Półka, M., et al., *Chapter The Use of Unmanned Aerial Vehicles by Urban Search and Rescue Groups.* 2018.

79. Murray, T; Hasan, S. F., *Received Signal Strength Indicator for Device Localization in Post Disaster Scenarios*, International Conference on Information Networking (ICOIN), 2022.

Chapter 7

Computational Mathematical Analysis of a Phytoplankton–Zooplankton Interactions Model

Rakesh Kumar[1] and Navneet Rana[2]
Deptt of Applied Sciences, S.B.S. Singh University, Ferozepur, Punjab, India

Abstract: In this chapter, the interactions of phytoplankton-zooplankton (a prey-predator interaction) with time delay are discussed and also explained its role in plankton ecology. The dynamics of both populations of the proposed system is studied. It is investigated that the parameter τ (time delay) can disturb the stability properties of the system and produced periodic oscillations in both populations due to Hopf-bifurcation. It has been shown that the increase in time delay caused complexities in the behaviour of populations. To verify the analytical results, a suitable example has been given in the numerical simulations for the justification of all the theoretical outputs.

1.1 Introduction

Plankton is one of the main components in the food chain system in water. Planktons are aquatic organisms. They are also named 'marine drifter'.

There are two types of plankton:-

1. **Plants known as Phytoplankton**:- They are plant-like substances. All the aquatic organisms depend on them for their food. They produce their food through the process of photosynthesis. Their growth in water is highly influenced by the available nutrients in the water.
2. **Animals known as Zooplankton**:- They are animal-like substances. Their main food is phytoplankton. There are two types of zooplankton, primary and secondary. Primary zooplanktons eat phytoplankton whereas secondary zooplanktons eat primary zooplankton.

Many mathematical models have suggested knowing the behaviour of Phytoplankton-Zooplankton system. These models studied how the Phytoplankton and Zooplankton populations are related to each other, how nutrients affect their growth, and how toxin substances in water vary their growth has been analysed by Chattopadhayay et al. [1]. The authors have shown the impact of water on the dynamics of phytoplankton and zooplankton populations [2]. A delayed differential equations system on harmful algae blooms in the presence of toxic substances has been given in [3]. Saha and Bandyopadhyay discussed the delayed system are studied to understand the behaviour of toxin-producing interactions of phytoplankton and zooplankton populations, and introduced Holling II function to study variation in the dynamic behaviour [4]. The behavioural analysis of a model with delay has been given to analyse the interactions between phytoplankton and zooplankton and observed Hopf bifurcation [5, 6]. A model with Holling IV type function has been discussed by researchers in [7] to present the interactions between phytoplankton-zooplankton which are producing toxin and found that time delay produced oscillations in population due to Hopf-bifurcation. Liao et al. [8] investigated the effect of noise in toxic phytoplankton and two zooplankton populations system with delay. They established the impact of noise on stochastic extinction. Also, they observed that the persistence of both populations is the same, but the synergistic effects of the noises on the stochastic extinction, and the persistence of the planktons are stronger than that of single noise. Further, an increase in the rate of liberation of the toxin of both zooplanktons was found to be able to increase the biomass of the phytoplankton, and minimise the zooplankton biomass. Many researchers have developed Mathematical systems, which are proven to be a powerful tool for examining the dynamics of aquatic plankton models in ecology qualitatively and quantitatively, and the important research outputs have been used to develop the growth mechanisms of zooplanktons as well as of phytoplankton, and are able to find some important factors responsible for creating phytoplankton blooms [9-11].

In this chapter, we are going to study the dynamics of phytoplankton-zooplankton system with time delay. The time delay is considered in zooplankton population caused by their migration in horizontal or vertical direction. The rest of the paper is organised as follows: In section 1.2, the mathematical model of the system is formulated. The positivity of the solutions and their boundedness is discussed in section 1.3. In section 1.4, the possibility of various equilibriums and their stability with and without delay has been discussed. The dynamical analysis along with Hopf Bifurcation has been discussed in section 1.5. After that in section 1.6, a numerical example to support the analytical outcomes has been presented. Finally, the outcomes of this mathematical model are discussed as conclusions in the last section.

1.2 The Mathematical Model

The model involves the interactions between two population densities namely Phytoplankton with density P(t) and Zooplankton with density Z(t). Let τ be the time delay in zooplankton predation, which arises due to time spent by zooplankton population in vertical and horizontal migration caused by the highest predators such as fish. The model is represented by system of equations (1): of the system.

$$\left.\begin{aligned}\frac{dP}{dt} &= rP - \delta P^2 - \frac{\beta P(t)Z(t-\tau)}{\gamma + P(t)}\\ \frac{dZ}{dt} &= \frac{\beta_1 \beta P(t)Z(t-\tau)}{\gamma + P(t)} - \delta_1 Z(t)\end{aligned}\right\} \qquad (1)$$

The initial condition for the system (1) is

$P(\theta) = \phi_1(\theta), Z(\theta) = \phi_2(\theta), \phi_1(\theta) \geq 0, \phi_2(\theta) \geq 0, \theta \in [-\tau, 0]$ and $\phi_1(\theta) > 0, \phi_2(\theta) > 0.$

Let $C([-\tau,0], R_+^2)$ be the Banach space of continuous functions defined on the interval $[-\tau, 0]$ into R_+^2 where

$R_+^2 = \{(x_1, x_2) : x_i > 0, i = 1, 2\}$.Then $\phi_1(\theta), \phi_2(\theta) \in C([-\tau,0], R_+^2)$

Table 1: Description of Parameters

Parameters	Description
r	Intrinsic growth rate of phytoplankton
δ	Natural mortality rate phytoplankton
β	Capture rate of zooplankton on phyto-plankton
β_1	Conversion rate of phytoplankton to zooplankton
δ_1	Natural mortality rate zooplankton
γ	Half saturation constant

Without any time delay, i.e., in the absence of τ, (1) can be rewritten:

$$\left.\begin{aligned}\frac{dP}{dt} &= rP-\delta P^2-\frac{\beta P(t)Z(t)}{\gamma+P(t)}\\ \frac{dZ}{dt} &= \frac{\beta_1\beta P(t)Z(t)}{\gamma+P(t)}-\delta_1 Z(t)\end{aligned}\right\} \quad (2)$$

In the approaching section, the boundedness and the positivity of all the solutions have been discussed in detail.

1.3 Positivity of Solutions and its Boundedness

In this section, we are finding the conditions such that the system (1) has positive as well as bounded solutions.

Lemma 1.1: With initial conditions $\phi_1(\theta)>0, \phi_2(\theta)>0$ defined on $[0, +\infty)$, the solutions of model system (1) will always be in the positive quadrant for all $t \geq 0$.

Proof: Assume that (P(t), Z(t)) is a possible solution of the model system (1), and which is taken together with initial conditions. The first equation of the model system (1) will be written as:

$$\frac{dP(t)}{P(t)}=\left(r-\delta P(t)-\frac{\beta Z(t-\tau)}{\gamma+P(t)}\right)dt$$

which implies that $$\frac{dP(t)}{P(t)} = \eta(P,Z)dt,$$

where $$\eta(P,Z) = \left(r - \delta P(t) - \frac{\beta Z(t-\tau)}{\gamma + P(t)} \right).$$

Integrate the equation in the region [0, t], we obtain

$$P(t) = \phi_1(0)e^{\int \eta(P,Z)dt} > 0.$$

Similarly from the second equation of the system (1)

$$\frac{dZ}{dt} \geq -\delta_1 Z(t).$$

Now integrate this equation in the region [0,t], we obtain

$$Z(t) \geq \phi_2(0)e^{-\delta_1 t} > 0.$$

Therefore, P(t)>0, Z(t)>0 for all positive times, i.e., $t \geq 0$.

Hence the proof.

Lemma 1.2: The solutions of the formulated model system (1) in the positive quadrant R_+^2 are always bounded uniformly.

Proof: Suppose that (P(t), Z(t)) is any solution of (1) with defined positive initial conditions. From the first equation of system (1), we have

$$\frac{dP}{dt} \leq P(r - \delta P),$$

By comparison, the result given in [12], we can always have

$$\lim_{t \to \infty} \sup\ P(t) \leq \frac{r}{\delta} = K.$$

Let $\Theta(t) = \beta_1 P + Z$, which is completely time-dependent. We can have

$$\frac{d\Theta}{dt}+k\Theta=2r\beta_1\mathrm{P}-\beta_1\delta P^2,$$

where $k=\min\{r,\delta_1\}$

$$\therefore\ \frac{d\Theta}{dt}+k\Theta\leq 2r\beta_1\mathrm{P}\text{ , for any k>0.}$$

By simpler computations, we obtain

$$\lim_{t\to\infty}\ \sup\ \mathrm{Z}(t)\leq\frac{2r^2\beta_1}{\delta k}.$$

Therefore, all the solutions of model (1) are ultimately bounded.

1.4 Equilibrium Points and their Stability Analysis

In this section, we are going to analyse different equilibrium points of the system (1). There are three feasible equilibrium points, which are discussed as follows:

(i) $E_0(0,0)$ is the trivial equilibrium point, which is always feasible.

(ii) $E_1(P,0)$ is the zooplankton free equilibrium point, where $P=\frac{r}{\delta}$, and always feasible.

(iii) $E_2(P^*,Z^*)$ is the coexisting equilibrium point, where

$$P^*=\frac{\delta_1\gamma}{\beta_1\beta-\delta_1}\quad\text{and}\quad Z^*=\frac{\beta_1\gamma}{(\beta_1\beta-\delta_1)^2}(r\beta_1\beta-r\delta_1-\delta_1^2\gamma)$$

Here, P^* and Z^* is positive provided

$$\beta_1\beta>\delta_1\ \text{ and }\ \beta_1\beta>\delta_1+\frac{\delta\delta_1\ \gamma}{r}$$

which gives

$$\beta_1\beta>\max\left\{\delta_1\ ,\delta_1+\frac{\delta\delta_1\ \gamma}{r}\right\}.$$

This gives $\beta_1\beta>\delta_1+\frac{\delta\delta_1\ \gamma}{r}$.

Next, let us investigate the stability properties of coexisting equilibrium point E_2 and we will show that through Hopf bifurcation, it can be disappeared by taking τ as bifurcation parameter. By the variational matrix of plankton system (time-delayed), we linearize the system (1) around E_2 to obtain the characteristic equation of the given system:

Firstly, we calculate the Jacobian matrix at the corresponding equilibrium point $E_2(P^*, Z^*)$ to observe the behaviour of the solution trajectories:

$$J_{E_2} = \begin{bmatrix} S_{11} & S_{12} \\ S_{21} & S_{22} \end{bmatrix}, \text{ where}$$

$$S_{11} = r - 2\delta P^* - \frac{\beta\gamma Z^*(t-\tau)}{(\gamma+P^*)^2}, \quad S_{12} = \frac{-\beta P^* e^{-\lambda\tau}}{\gamma+P^*},$$

$$S_{21} = \frac{\beta_1\beta\gamma P^* Z^*(t-\tau)}{\left(\gamma+P^*\right)^2}, \quad S_{22} = \frac{\beta_1\beta P^* e^{-\lambda\tau}}{\gamma+P^*} - \delta_1.$$

The characteristic equation of J_{E_2} is

$$\Delta(\lambda,\tau)\big|_{E_2} = (\lambda^2 + A_1\lambda + A_2) + (A_3\lambda + A_4)e^{-\lambda\tau} = 0, \tag{3}$$

Where $A_1 = \delta_1 - r + 2\delta P^* + \dfrac{\beta\gamma Z^*(t-\tau)}{(\gamma+P^*)^2}$,

$$A_2 = \delta_1\left(-r + 2\delta P^* + \frac{\beta\gamma Z^*(t-\tau)}{(\gamma+P^*)^2}\right),$$

$$A_3 = \frac{-\beta_1\beta P^*}{\gamma+P^*}, \text{ and } A_4 = \frac{\beta_1\beta P^*}{\gamma+P^*}(r - 2\delta P^*).$$

The values A_1, A_2, A_3, A_4 are dependent on τ and all are differentiable functions from the interval [0, +∞) to the positive quadrant R_+^2, which are continuous as well.

1.5 Dynamical Analysis of the Model System

In the current section, model (1) is analysed qualitatively for the dynamical behaviour.

1.5.1 Model without Time Delay

In this subsection, we are going to analyse the system without any delay, i.e., $\tau = 0$.
For $\tau = 0$, the characteristic equation (3) becomes

$$\Delta(\lambda,\tau)\big|_{E_2} = \lambda^2 + (A_1 + A_3)\lambda + (A_2 + A_4) = 0 \tag{4}$$

where the value of A_1, A_2, A_3, A_4 are defined as above.
The Routh-Hurwitz criterion given in [13] for a quadratic polynomial derived from a two-dimensional system can be used to discuss the stability properties of (1). By this, we noticed that the system is asymptotically stable if all the values of the quadratic characteristic equation will have real parts with negative value, i.e., if

$$(H_1): A_1 + A_3 > 0 \quad \text{and} \quad A_2 + A_4 > 0$$

$$\text{i.e. } \frac{\delta\delta_1\gamma}{(\beta\beta_1 - \delta_1)} < r < \frac{\delta\gamma\beta_1\beta + \delta_1\delta\gamma}{(\beta\beta_1 - \delta_1)}. \tag{5}$$

Further, we can state the lemma as follows:-

Lemma 1.3: The coexisting equilibrium $E_2(P^*, Z^*)$ without any delay is locally asymptotically stable if the condition $\frac{\delta\delta_1\gamma}{(\beta\beta_1 - \delta_1)} < r < \frac{\delta\gamma\beta_1\beta + \delta_1\delta\gamma}{(\beta\beta_1 - \delta_1)}$ is satisfied.

1.5.2 Behaviour of the Delayed Model System

In this subsection, we are going to study the stability condition of the formulated model system (1) with nonnegative time delay ($\tau > 0$) around equilibrium point $E_2(P^*, Z^*)$ i.e., we are going to see how the real part of a value of equation (3) vanishes and becomes nonnegative with the variations in time delay τ. As a result of this, the Hopf bifurcation arises in the system (1).

The positive equilibrium $E_2(P^*, Z^*)$ is asymptotically stable if all the roots of the transcendental characteristic equation (3) have negative real parts. At τ = 0, all the solutions of the transcendental characteristic equation (3) have negative real parts if conditions (H_1) holds. Hence for the locally stability of model system (1), we hope that the conditions (H_1) hold well. By the continuous nature of the values $\lambda(\tau)$ of $\Delta(\lambda, \tau)$, the characteristic equation (3) can have values with positive real parts iff $\Delta(\lambda, \tau)$ has values with real part vanishes for some τ > 0, i.e., the nature of all the roots of characteristic equation (3) will be purely imaginary. The probability for the coexisting steady state $E_2(P^*, Z^*)$ to become unstable is $\Delta(0, \tau) = 0$ for some positive τ or that $\Delta(\lambda, \tau) = 0$ has a conjugate pair of purely imaginary values for positive value of τ. From the characteristic equation (3), it is obvious that $\Delta(0, \tau) = A_2 + A_4 \neq 0$. So, the only possibility for the coexisting steady state to turn to be unstable is that a purely imaginary pair of values exists at the threshold value $\tau_0^* > 0$.

Also, if τ_0^* is the smallest root at which a purely imaginary conjugate pair of values exist, then the other eigenvalues of (3) must have negative real parts $\tau = \tau_0^*$.

On the basis of this result, we can state the lemma as:-

Theorem 1.4: [14] (i) The coexisting point $E_2(P^*, Z^*)$ of model (1) is locally stable iff $E_2(P^*, Z^*)$ of model system (1) is asymptotically stable, and the transcendental equation (3) does not have complex roots for positive τ.

(ii) The coexisting point $E_2(P^*, Z^*)$ of model (1) is conditionally stable if all values of transcendental equation (3) does not have the positive real part for τ = 0, and ∃ nonnegative value of τ so that (3) have a pair of imaginary values $\pm i\omega$.

It is obvious that the coexisting point $E_2(P^*, Z^*)$ is locally stable if transcendental equation (3) has roots $\mathrm{Re}(\lambda) < 0$. We can state the following theorem in which the nonexistence of instability with the enhancement τ is given by Gopalsamy [15].

Theorem 1.5: The necessary and sufficient condition for coexisting point $E_2(P^*, Z^*)$ to be stable (locally) with the involvement of τ is:

1. The real part of i.e. Re(all the roots of $\Delta(\lambda,\tau)=0$) is negative.
2. For all real ω and $\tau > 0$, $\Delta(\lambda,\tau) \neq 0$

Now we are going to find the condition for the existence of Hopf bifurcation around coexisting point E_2 in a model τ. Using τ as a control parameter for Hopf-bifurcation.

Theorem 1.6: The coexisting equilibrium point $E_2(P^*, Z^*)$ is locally asymptotically stable for any time delay if (H_2) holds, where $(H_2): A_1^2 - A_3^2 - 2A_2 > 0, A_2^2 - A_4^2 > 0$.

Proof: Here, we are going to find a complex pair of values of the transcendental equation (3).

Let, for a $\tau > 0, \lambda = i\omega$ ($\omega > 0$ and $i = \sqrt{-1}$)is a value of transcendental equation (3), ω is positive real. Put $\lambda = i\omega$ in (3) and using the techniques from [16] for finding real and imaginary parts of transcendental equation (3), we can obtain

$$A_4 \cos\omega\tau + A_3\omega \sin\omega\tau = \omega^2 - A_2, \tag{6}$$

$$A_3\omega\cos\omega\tau - A_4 \sin\omega\tau = -A_1\omega. \tag{7}$$

By using simple computational techniques for simplifying (6) and (7), we obtain 4th order equation in ω as

$$\omega^4 + (A_1^2 - A_3^2 - 2A_2)\omega^2 + (A_2^2 - A_4^2) = 0, \tag{8}$$

It will be expressed as

$$B(z) = z^2 + B_1 z + B_2 = 0\,, \qquad (*)$$

where $z = \omega^2$ and $B_1 = A_1^2 - A_3^2 - 2A_2$ and $B_2 = A_2^2 - A_4^2$

From the above equation if

$(H_2): A_1^2 - A_3^2 - 2A_2 > 0$ and $A_2^2 - A_4^2 > 0$.

Therefore, the system (*) does not have positive roots if

$A_1^2 - A_3^2 - 2A_2 > 0$ and $A_2^2 - A_4^2 > 0$.

We observed from the condition that if (H_1) and (H_2) holds, and then all values of (4) have $\text{Re}(\lambda) < 0$. It follows by Rouche's results that the values of ω in (8) have $\text{Re}(\omega) < 0$. Thus, the coexisting equilibrium E_2 is locally stable for any time delay $\tau > 0$.

1.5.3 Hopf Bifurcations

Now we are going to find an imaginary pair of values of transcendental equation. (3).
If $A_2^2 - A_4^2 < 0$, then by Routh Hurwitz phenomenon, equationn (8) must have a nonnegative root $\omega_0^{\,2}$, which is unique positive. This gives us that equation (3) must contain a pair of imaginary values $\pm i\omega_0$, this will leads to Hopf bifurcation in model system (1).
Put $\omega = \omega_0^{\,2}$ in (6) and (7), simplifying to get τ. The value τ is

$$\tau_n^* = \frac{1}{\omega_0}\sin^{-1}\left\{\frac{\omega_0\left(\omega_0^{\,2}A_3 - A_2A_3 + A_1A_4\right)}{A_3^{\,2}\omega_0^{\,2} + A_4^{\,2}}\right\} + \frac{2n\pi}{\omega_0};n = 0,1,2,3,.... \tag{9}$$

By Butler Lemma [17], the positive coexisting point $E_2(P^*, Z^*)$ will be stable when $\tau < \tau_0^*$ for $n = 0$.
Further, we inspect model (1) for Hopf bifurcation at $E_2(P^*, Z^*)$ when τ increases through the critical value τ_0^*. As given in the theory of bifurcation in [18, 19], two results for Hopf bifurcation stated as:

(i) A complex pair of eigenvalues $\pm i\omega_0$ exists, and remaining eigenvalues with $\text{Re}(\lambda) < 0$, which has already been stated and proved in Theorem 1.4.

(ii) The transversality condition is satisfied.

Next we will check the condition of transversality for Hopf bifurcation to investigate about the variation in real part of characteristic equation (3) with τ near τ_0^* i.e.

$$\left\{\frac{d}{d\tau}(\operatorname{Re}\lambda)\right\}_{\tau=\tau_0^*,\omega=\omega_0} > 0$$

The condition of Hopf bifurcation is satisfied giving the required periodic solution. Let $\lambda(\tau)=\mu(\tau)+i\omega(\tau)$ be complex root of characteristic equation (3) near $\tau=\tau_0^*$, such that $\mu(\tau_0^*)=0$ and $\omega(\tau_0^*)=\omega_0$. Substituting $\lambda(\tau)$ into (3), differentiate with τ, we get

$$sign\left\{\frac{d}{d\tau}(\operatorname{Re}\lambda)\right\}_{\tau=\tau_0^*,\omega=\omega_0} = sign\left\{\frac{\sqrt{(A_3^2-A_1^2+2A_2)^2-4(A_2^2-A_4^2)}}{((-\omega_0^2+A_2)^2+A_1^2\omega_0^2)(A_4^2+A_3^2\omega_0^2)}\right\} \tag{10}$$

By virtue of (H_3), $\left\{\frac{d}{d\tau}(\operatorname{Re}\lambda)\right\}_{\tau=\tau_0^*,\omega=\omega_0} > 0$.

Thus the transversality result holds good.
Based on this analysis, another Theorem to discuss conditions can be stated as:

Theorem 1.7: [20]
Assume that $E_2(P^*,Z^*)$ exists and the conditions $A_1^2-A_3^2-2A_2>0, A_2^2-A_4^2<0$ justified for the model system (1). The necessary and sufficient conditions that the coexisting point $E_2(P^*,Z^*)$ is L.A.S. with the involvement of time delay are

(i) The coexisting equilibrium $E_2(P^*,Z^*)$ is locally asymptotically stable $0\le\tau<\tau_0^*$.

(ii) The coexisting equilibrium $E_2(P^*,Z^*)$ becomes unstable $\tau>\tau_0^*$.

(iii) For *n* =0, the equation (9) becomes $\tau_0^*=\frac{1}{\omega_0}\sin^{-1}\left\{\frac{\omega_0\left(\omega_0^2A_3-A_2A_3+A_1A_4\right)}{A_3^2\omega_0^2+A_4^2}\right\}$, and the equations (1) proved to have Hopf bifurcation for at threshold value $\tau=\tau_0^*$ about E_2.

1.6 Numerical Simulations

This section deals with the dynamical behaviour of the system by numerical computations. Let us take a set of values for distinct parametric as $r = 0.25, \beta = 1.0, \beta_1 = 0.67, \gamma = 1, \delta_1 = 0.2, \delta = 0.2$ for the system (1), i.e.,

$$\left.\begin{aligned} \frac{dP}{dt} &= (0.25)P(t) - (0.2)P^2 - \frac{(1.0)P(t)Z(t-\tau)}{1+P(t)} \\ \frac{dZ}{dt} &= \frac{(0.67)(1.0)P(t)Z(t-\tau)}{1+P(t)} - (0.2)Z(t) \end{aligned}\right\} \quad (11)$$

By using $P(0) = 0.31$ and $Z(0) = 0.1$ for the example (11) and $\tau = 0$, (11) converges to stable coexisting point $E_2(0.4264, 0.2342)$ which is as shown in fig. 1.

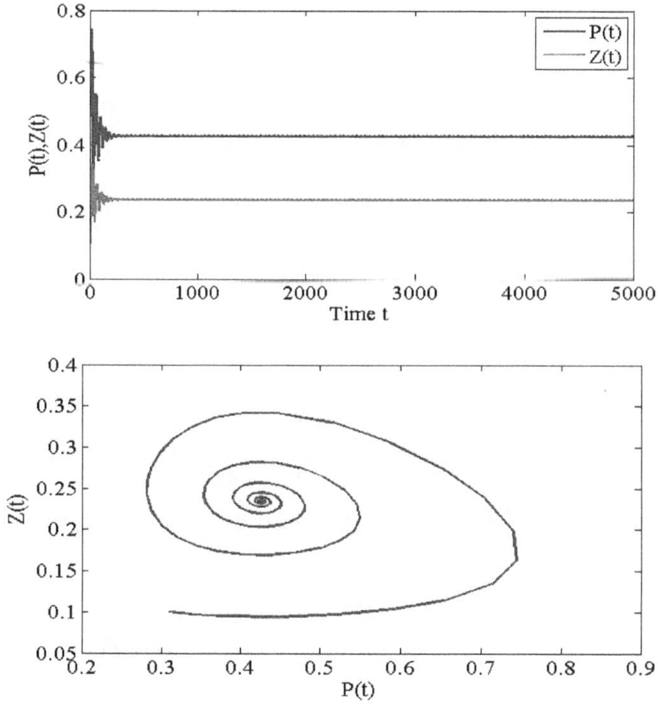

Fig.1. Solution curves proving the locally stable behaviour for phytoplankton and zooplankton population for $\tau = 0$.

The system is integrated with delay τ by MATLAB and attains stability at $\tau = 1.2$ and this stable dynamical behaviour is shown in fig.2.

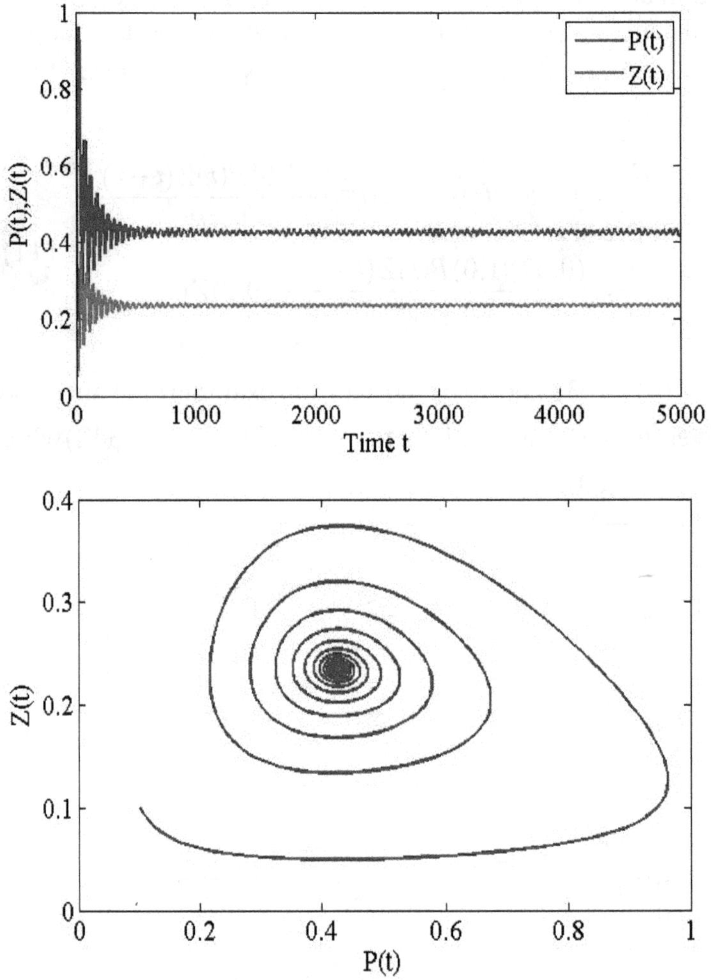

Fig.2. Convergence of solution trajectories to $E_2(0.4264, 0.2342)$, of the system (1) at $\tau = 1.2 < \tau_0^*$.

But with a little enhancement in τ by fixing all the parameters, we have found that small orbits occur in the system and Hopf bifurcation came to existence in the system. The oscillatory behaviour of phytoplankton and zooplankton populations at $\tau_0^* = 3.35$ around the positive equilibrium point E_2 is shown in fig.3.

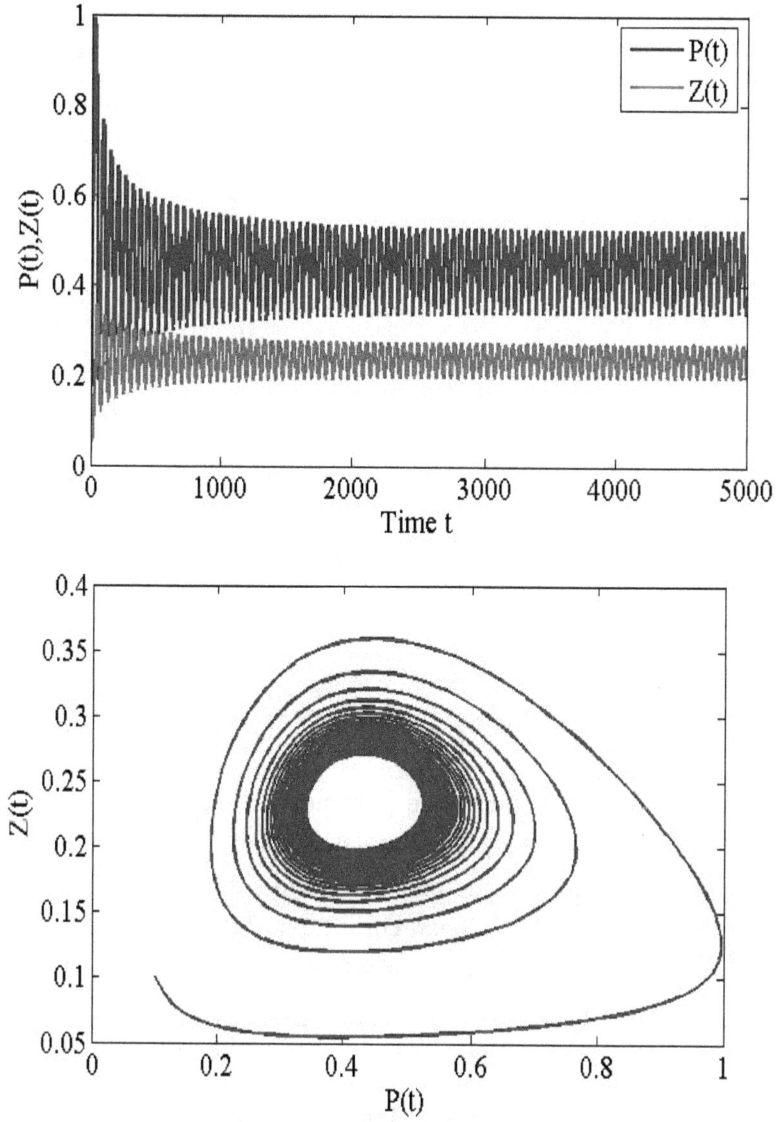

Fig.3. Oscillatory Behavior of phytoplankton and zooplankton system at $\tau_0^* = 3.35$ around the equilibrium point E_2.

The stable periodic solutions $\tau = 15.2$ is demonstrated in fig.4 for model (11).

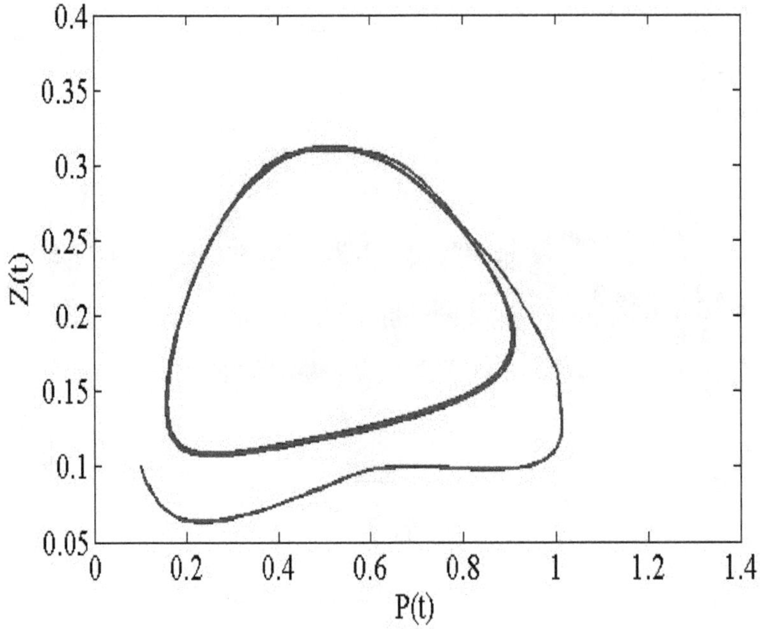

Fig.4. Stable periodic solution around E_2 for phytoplankton-zooplankton population at $\tau_0^* = 15.2$.

Numerically, by using the above parametric values, the conditions $A_1^2 - A_3^2 - 2A_2 > 0, A_2^2 - A_4^2 < 0$ are satisfied. By Routh-Hurwitz results, $\exists$ a nonnegative value of ω (8), and thus an imaginary root $i\omega_0$, with $\omega_0 = 0.8134$ has been computed. Putting $\omega_0 = 0.8134$ in (9), we have $\tau_0^* = 3.35$, which is the critical value of time delay for the system such that when the system crosses this critical value, it loses its stability. Moreover, the transversality condition $\left\{\frac{d}{d\tau}(\operatorname{Re}\lambda)\right\}_{\tau=\tau_0^*,\omega=\omega_0}$ 0.277834 > 0 has been verified, which is the primary result for the occurrence of Hopf bifurcation. The complex behaviour in the system has been produced at $\tau = 35.5$ is shown in fig.5.

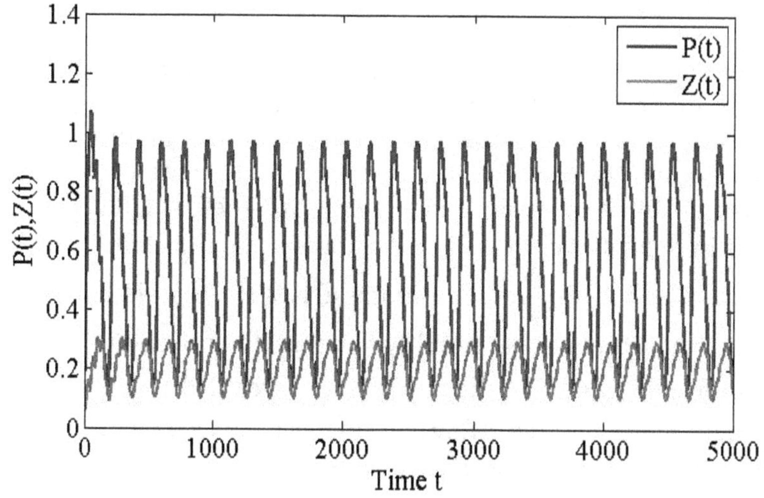

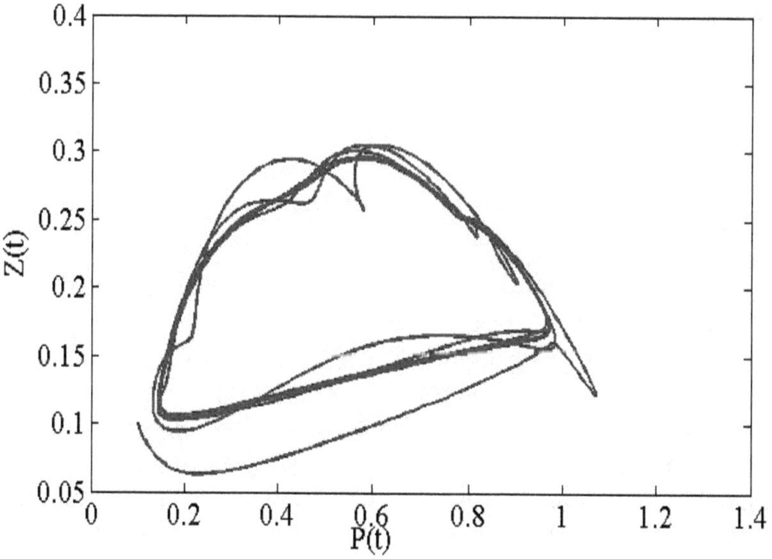

Fig.5. Complex behaviour has been produced by system at $\tau = 35.5$.

1.7 Conclusions

In this chapter, we have investigated the dynamics of the time-delayed model of phytoplankton-zooplankton system. The preliminary results containing the positivity of various solutions of the model (1) and the uniform boundedness of the solutions have been discussed. Different equilibrium points and their stability are discussed in details. We noticed that the trivial equilibrium E_0 always feasible. The zooplankton free point E_1 is also feasible and satisfy certain condition for local stability. Further, the coexisting point E_2 is also feasible and is L.A.S. up to the threshold value of τ. The threshold value of time delay parameter $\tau_0^* = 3.35$ has been computed which is responsible for Hopf bifurcation in the model (1). The primary results for the occurrence of Hopf bifurcation have also been verified. Also, with the increase in delay parameter we have been able to observe complex situations in the system. The complex dynamical behaviour of model (1) has been depicted in Fig. 5.

References

1. Chattopadhayay, J., Sarkar, R.R., and Mandal, S. (2002) Toxin-producing plankton may act as a biological control for planktonic blooms field study and mathematical modelling. *Journal of Theoretical Biology*, **215**(3), 333–344.
2. Article: Zhao, Q., Liu, S., and Niu, X. (2020) Effect of water temperature on the dynamic behavior of phytoplankton--zooplankton model. *Applied Mathematics and Computation, Elsevier*, **378**, 125211.
3. Article: Chattopadhyay, J., Sarkar, R.R., and El Abdllaoui, A. (2002) A delay differential equation model on harmful algal blooms in the presence of toxic substances. *Mathematical Medicine and Biology: A Journal of the IMA*, **19**(2), 137–161.
4. Article: Saha, T., and Bandyopadhyay, M. (2009) Dynamical analysis of toxin producing phytoplankton–zooplankton interactions. *Nonlinear Analysis: Real World Applications*, **10**(1), 314–332.

5. Article: Rehim, M., and Imran, M. (2012) Dynamical analysis of a delay model of phytoplankton–zooplankton interaction. *Applied Mathematical Modelling,* **36**(2), 638–647.
6. Article: Jiang, Z., Ma, W., and Li, D. (2014) Dynamical behavior of a delay differential equation system on toxin producing phytoplankton and zooplankton interaction. *Japan Journal of Industrial and Applied Mathematics, Springer,* **31**, 583-609.
7. Article: Sharma, A., Sharma, A. K., and Agnihotri, K. (2015) Analysis of a toxin producing phytoplankton-zooplankton interaction with Holling IV type scheme and time delay. *Nonlinear Dynamics, Springer,* **81**, 13-25.
8. Article: Tiancai, L., Hengguo, Yu., Chuanjun, D., Min, Z. (2020) Impact of noise in a phytoplankton-zooplankton system. *Journal of Applied Analysis & Computation,* **10**(5), 1878-1896.
9. Article: Chaudhuri, S., Roy, S., & Chattopadhyay, J. (2013) Phytoplankton–zooplankton dynamics in the 'presence'or 'absence'of toxic phytoplankton. *Applied Mathematics and Computation,* ***225***, 102-116.
10. Article: Dai, C., Zhao, M., Yu, H., & Wang, Y. (2015) Delay-induced instability in a nutrient-phytoplankton system with flow. *Physical Review,* **91**(3), 032929.
11. Article: Jang, S., Baglama, J., & Wu, L. (2014) Dynamics of phytoplankton–zooplankton systems with toxin producing phytoplankton. *Applied Mathematics and Computation,* **227**, 717-740.
12. Book: Birkhoff, G., and Rota, G. (1989) *Ordinary differential equations.* Ginn, Boston.
13. Book: Luenberger, D.G.D.G. (1979) *Introduction to dynamic systems; theory, models, and applications,* **4**; QA402, L8.
14. Aricle: Boonrangsiman, S., Bunwong, K., and Moore, E.J. (2016) A bifurcation path to chaos in a time-delay fisheries predator–prey model with prey consumption by immature and mature predators. *Mathematics and Computers in Simulation,* **124**, 16–29.
15. Book: Gopalsamy, K. (2013) *Stability and oscillations in delay differential equations of population dynamics,* **74**, Springer Science & Business Media.
16. Article: Kumar, R. (2021) 10 Mathematical modeling of a delayed innovation diffusion model with media coverage in adoption of an innovation. In *Systems Reliability Engineering,* De Gruyter, pp. 153-172.

17. Article: Freedman, H., and Waltman, P. (1985) Persistence in a model of three competitive populations. *Mathematical Biosciences,* **73**(1), 89–101.
18. Book: Edelstein-Keshet, L. (1988) *Mathematical models in biology,* **46**. Siam.
19. Book: Kuznetsov, Y.A. (2004) *Elements of Applied Bifurcation Theory,* Third Edition (Applied Mathematical Sciences), **112**, New York: Springer Verlag.
20. Article: Kumar, R., Sharma, A. K., and Agnihotri, K. (2019) Bifurcation analysis of a nonlinear diffusion model: Effect of evaluation period for the diffusion of a technology. *Arab Journal of Mathematical Sciences,* **25**(2), 189-213.

Chapter 8

Association between Socio-Economic and Demographic Factors Influencing the Present Level of Financial Literacy about Variable Income Investments

Vijay Laxmi, S.S.D. Women's Institute of Technology, Bathinda, India
Kiran Bala, Mata Sundri University Girls College, Mansa, India

1. Introduction

Financial literacy is that set of skills which help the investors in understanding the financial information, financial product and concepts to take appropriate financial decisions. It is concerned with Knowledge for better money management practices concerned with saving, spending, investing and borrowing and developing skills and confidence to make good financial choices. As per the reports of various surveys, low financial literacy is found around the world. In case of Indian states and on the basis of Gender, lower financial literacy is measured among females than males. Here, researcher has made an attempt to find the financial awareness, association between financial literacy and various demographic and socio-economic factors and the factors influencing present level of financial literacy among working women in Punjab. Figures show that majority of women respondents possess financial knowledge as they know about the investment modes but lack financial literacy as few of them have invested in these modes. Chi-square test shows significant association between financial literacy and socio-economic and demographic factors undertaken for this study. Results of multiple logistic regression shows that level of education, monthly income and women engaged in manufacturing sector are the factors influencing the present level of financial literacy among working women in Punjab.

1.1 Need and significance of financial literacy

The concept of financial literacy has gained importance and presently it has become a major issue in the rapidly growing India due to the inability of women to understand the financial concepts and their lack of financial knowledge. Knowledge of managing personal financial resources by all human beings is the basic aim to study financial literacy. One's ability to understand, compare and utilize knowledge to make investments in actual practice is termed as financial literacy. But it's not an easy task for each and every one to manage efficiently her funds without possessing financial knowledge and skills. Henceforth, financial education is not only important for men but it's of paramount importance to women too. It will result in better understanding and knowledge of complexities of rapidly growing financial products and improve the skills and literacy for informed financial choices. It will help women to become independent financially. Financial education is important for women because their life expectancy is higher than men and live longer than them. Thus, their financial needs would rise and require more savings.

According to Moore (2003) [1], Individuals are considered financially literate if they are competent and can demonstrate they have used the knowledge they have learned. Literacy is obtained through practical experience and active integration of knowledge."

2. Literature review

Author	Objective	Sample size	Variables	Statistical tools	Findings
Bhushan & Medury (2013)	To measure the financial literacy	516 salaried individuals in Himachal Pradesh	Education level, level of income, workplace and nature of employment &geographical region	Correlation	High level of financial literacy among males than females. Education level, level of income, workplace

					and nature of employment positively influenced financial literacy. Dependence of FL on education level was found
Jain (2014)	To analyze the pattern and relationship of income and investment	250 working women from govt. sector & private sectors of Ahmadabad city	Age, Education Level, Monthly Income and work sectors	Percentage, Mean and Std. deviation	Income level influenced the level of savings. They preferred to invest in Bank fixed deposits followed by gold &insurance and gave equal importance to PPF, Post office

					schemes and real estate
Chatterjee, k. (2018)	To assess the level for basic and advanced financial literacy and to find the difference in levels of financial literacy with changes in various socio-economic and demographic variables.	200 adults from an urban agglomeration of Kolkata named Bidhanagar township of India.	Age, education, occupation, marital status, gender, family system, role in financial decision making, the contribution of income by family members.	Descriptive statistics, coefficient of variance & Chi-square	A highly significant relation was found between basic and advance financial literacy with age, gender and monthly income.

3. Objectives

1. To find out the financial awareness about modes of investment having a variable income.
2. To find out the association between financial literacy and various demographic and socio-economic factors.
3. To determine the factors influencing the present level of financial literacy among working women in Punjab.

4. Research Methodology

- Type of research: Descriptive
- Sample Size: 500

- Target Population: Employed and self-employed women working in Punjab
- Sampling Technique: Stratified random sampling
- Data collection: By questionnaire

5. Results and Discussion

Objective 1:

Table 5.1 Financial Awareness about Modes of Investment having Variable Income

Investment Options	Aware (Financial Knowledge)		Users (Financial Literacy)		Not Aware (No awareness)	
	F	%	F	%	F	%
Equity Shares	297	59.4	92	18.4	203	40.6
Investments in Commodity Market	223	44.6	37	7.4	277	55.4
Mutual Funds/ systematic investment plan	382	76.4	217	43.4	118	23.6
Real Estate / Property	402	80.4	177	35.4	98	19.6
Gold/ Silver (bullion)	334	66.8	142	28.4	166	33.2
Life Insurance	472	94.4	354	70.8	28	5.6

Source: Primary Survey conducted by the researcher

Financial knowledge, awareness and literacy for each individual investment mode carrying variable income have been observed from the above table. The majority of working women have financial knowledge about life insurance (94.4%) followed by investment in real estate (80.4%), mutual fund (76.4%), gold/silver (bullion) 66.8%, equity shares (59.4%) and commodity markets (44.6%) whereas financial literacy is found for life insurance among 70.8% respondents followed by mutual funds 43.4%, real estate 35.4%, gold/silver (bullion) 28.4% equity shares 18.4% and commodity market investments 7.4%. It has been concluded from the above analysis that women prefer to invest in low or moderate risk carrying investment options. Life insurance and mutual funds are more preferred than high risk carrying securities i.e. equity share and commodity market investments. Equity shares and commodity markets are the least used options. Financial literacy is identified in less number of respondents for both of these investment alternatives as compared to other investment modes.

Higher FL has been identified among women for life insurance followed by mutual funds, real estate and gold/silver. Although a large number of working women are financially knowledgeable about all the investments a wider gap persists between knowledge and financial literacy because few of them make use of knowledge in practice.

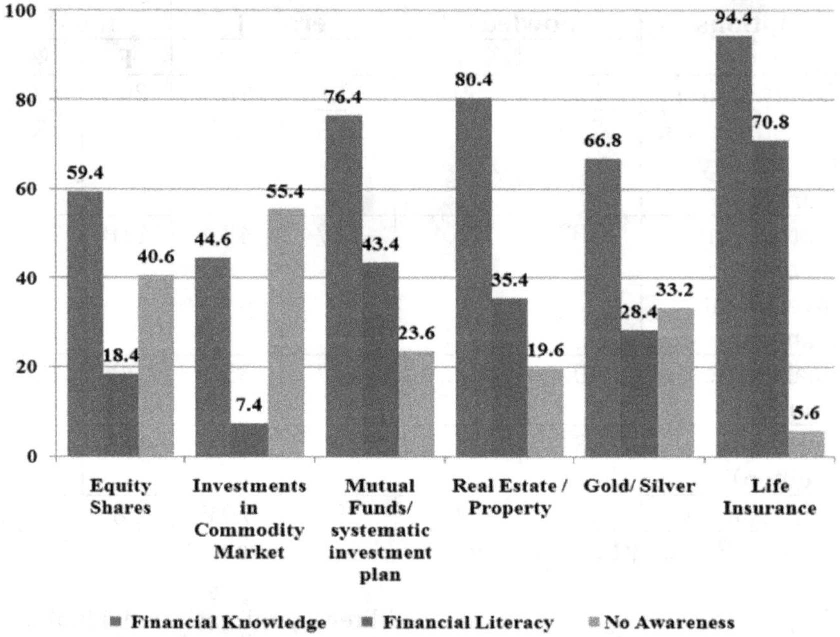

Figure 5.1 - Financial Knowledge, Literacy and Awareness for Individual Investment Mode

Source: Primary data by author

Above figure 5.1 exhibits that maximum financial knowledge is seen for investment in life insurance schemes followed by real estate, mutual funds, gold and silver, investments in equity shares and commodity market. Financial literacy for life insurance can be seen among more respondents than for financial literacy for mutual funds. Financial literacy is seen least for investments in the commodity market.

Objective 2:

5.2.1 Null Hypothesis Statements for Association between the Socio-Economic & Demographic Variables and Financial Literacy for Variable Income Investments

$H_0$1: There is no significant association between financial literacy level for variable income investment options and the age of women.
$H_0$2: There is no significant association between financial literacy level for variable income investment options and the level of education of women.
$H_0$3: There is no significant association between financial literacy level for variable income investment options and monthly income of women.
$H_0$4: There is no significant association between financial literacy level for variable income investment options and work sectors.
$H_0$5: There is no significant association between financial literacy level for variable income investment options and work experience of women.
$H_0$6: There is no significant association between financial literacy level for variable income investment options and marital status of women.
$H_0$7: There is no significant association between financial literacy level for variable income investment options and type of family of women.
$H_0$8: There is no significant association between financial literacy level for variable income investment options and financial decision making by women.
$H_0$9: There is no significant association between financial literacy level for variable income investment options and city.

5.2.2 Results of Chi-Square Performed to find Socio-Economic & Demographic Variables Influencing Variable Income Investment Financial Literacy

From the null hypothesis statements from **$H_0$1 to $H_0$9** pertaining to the **association between financial literacy level for variable in-**

come investment options and Socio-Economic & Demographic Variables Influencing variable Income Investment FL i.e. Age, level of education, income, work sectors, work experience, marital status, type of family, financial decision making and the city the following results have been observed. The value of Chi-square is found to be significant at 0.01 (**)/ 0.05*.

For age** (37.9), education level** (59.8), income** (97.8), work sectors** (72.3), work experience **(41.5), marital status **(41.5), type of family** (7.20), financial decision making** (12.9) and cities *(18.5).

Hence all the null hypothesis statements assuming 'no significant association' are rejected as the significant association is found for Age, level of education, monthly income, work sectors, work experience, marital status, type of family, financial decision making and city.

Objective 3:

5.3.1 Multiple Logistic Regression Model for Variable income investment FL:

log(y/1-y) = a+b_1 Type of family +b_2 Financial decision making +b_3 Age+b_4Monthly Income + b_5 Marital Status+b_6 Level of Education+b_7Years of Work Experience+b_8Work Sectors+ b_9Cities I

Independent variables = Different categories of Socio-economic and demographic variables with significant association. Dependent variable = Dichotomous categories of present level of financial literacy. Impact is categorized in two groups:

(I) High financial literacy
(II) Low financial literacy

Table 5. 2 Present Level of Financial Literacy among Working Women

Type of investments	Present Level of Financial Literacy			Total
	More n(%)	Less n(%)	No n(%)	
Financial Literacy Level for Variable	173(34.6)	238(47.6)	89(17.8)	500(100)

Income Investment				

Source: Primary Survey conducted by the researcher

'More' financial literacy is observed among 34.6% of respondents while 47.6% of respondents have 'Less' financial literacy for variable income investment modes. 17.8% of respondents have never invested even in a single variable income investment option. It means the financial literacy of such respondents for variable income carrying modes of investments is zero.

Table 5.3 Results of Multiple Logistic Regression for Financial Literacy of Variable Income Investment Options

		Beta co-eff.	S.E.	Wald Stats.	df	p-value	Odd ratios	95% C.I for odd ratio	
								Lower	Upper
Type of Family	Nuclear (RF)	-	-	-	-	-	-	-	-
	Joint	0.04	0.23	0.03	1	0.87	1.04	0.66	1.63
Financial Decision Making	Self	0.42	0.24	3.12	1	0.08	1.52	0.96	2.41
	Anyone else (RF)	-	-	-	-	-	-	-	-
Age	18-24yrs	-0.81	0.67	1.49	1	0.22	0.44	0.12	1.64
	25-34yrs	-0.26	0.45	0.33	1	0.57	0.77	0.32	1.86
	35-44yrs	0.03	0.38	0.01	1	0.94	1.03	0.48	2.19
	>44yrs	-	-	-	-	-	-	-	-
Monthly Income	Up to Rs. 20,000 (RF)	-	-	-	-	-	-	-	-
	Rs. 20001-40,000	0.80	0.43	3.57	1	0.06	2.23	0.97	5.14
	Rs. 40,001-60,000	0.84	0.44	3.63	1	0.06	2.32	0.98	5.49
	>Rs. 60,000	1.68	0.48	12.18	1	0.00**	5.36	2.09	13.76
Marital status	Married	0.96	0.81	1.40	1	0.24	2.61	0.53	12.75
	Unmarried	0.79	0.85	0.85	1	0.36	2.20	0.41	11.72
	Widow/Divorce (RF)	-	-	-	-	-	-	-	-
Level of Education	Graduation and below	-	-	-	-	-	-	-	-
	PG and other	0.59	0.25	5.47	1	0.02*	1.80	1.10	2.94
Year of work experience	Up to 5yrs (RF)	-	-	-	-	-	-	-	-
	6-15yrs	-0.02	0.30	0.00	1	0.95	0.98	0.55	1.75
	16-30yrs	-	0.5	1.18	1	0.28	0.57	0.21	1.57

		Beta co-eff.	S.E.	Wald Stats.	df	p-value	Odd ratios	95% C.I for odd ratio	
								Lower	Upper
		0.56	2						
	30 yrs or more	-0.17	0.72	0.05	1	0.82	0.85	0.21	3.47
Work sectors	Banking	-	-	-	-	-	-	-	-
	Education	-0.43	0.34	1.53	1	0.19	0.65	0.33	1.28
	Independent Professional	-0.44	0.35	1.52	1	0.24	0.65	0.32	1.329
	Insurance	-0.42	0.32	1.79	1	0.20	0.66	0.35	1.22
	Manufacturing	-1.79	0.47	14.25	1	0.00**	0.17	0.07	0.42
Cities	Amritsar(RF)	-	-	-	-	-	-	-	-
	Bathinda	0.67	0.34	3.83	1	0.05	1.96	0.99	3.83
	Jalandhar	-0.22	0.35	0.38	1	0.54	0.80	0.40	1.61
	Ludhiana	0.38	0.35	1.18	1	0.28	1.46	0.74	2.88
	Patiala	-0.29	0.36	0.66	1	0.42	0.75	0.37	1.51
	Constant	-2.44	1.08	5.11	1	0.02	0.09		

- Multiple logistic regression has been applied to find the impact of socio-economic and demographic variables on the present level of financial literacy. The estimated value of Beta coefficients is found positive for different categories of independent variable 'monthly income'. P-value is found statistically significant at 1% level of significance for Monthly income 'more than 60,000'. Table 5.3 shows an odd ratio of 5.36 which is more than 1. It predicts that the level of financial literacy will be higher as compared to income category 'up to Rs. 20,000' (reference category) when one unit of an independent variable is increased.
- Beta estimates for variable 'Level of education is positive for category 'PG and others' and p-value=0.02 is found significant at 5% level of significance. The odd ratio is more than 1 for 'post-graduation and others' and exhibits more chances of the higher level of FL than 'graduation and below' (RF).
- Women engaged in the manufacturing sector has a significant p-value = 0.00 at 1% level of significance. But the negative coefficient value (-1.79) and less than one odd ratio (0.17) predicts a lower level of financial literacy among women in the manufac-

turing sector as compared to women working in the banking sector (as RF).

- Other variables such as age, years of work experience, type of family, financial decision making, cities and marital status have no statistical significance at 5%/ 1% level of significance. Hence, it is concluded from the above results that these variables have no significance in influencing the level of present financial literacy about variable income bearing modes of investment.

6. Conclusions & Recommendations

The present study demonstrates that women prefer moderate risk-bearing investments like mutual funds and insurance schemes to meet future short term financial requirements. Levels of education, monthly income, women working in the manufacturing sector are prominent variables that influence their variable income investment financial literacy. Financial literacy is necessary for women to bring changes in their thinking, attitude and behaviour towards new financial products. It would not only improve their standard of living but also empower them financially and will enhance the development of an economy. Government and policymakers should plan out the different ways to spread financial literacy for modern and variable income bearing investment modes among women as the present level of financial literacy for variable income investments are not very high among working women. Seminars, campaigns and workshops must be organised to disseminate financial literacy and improve its level from time to time.

References

1. Agarwalla, Barua, Jacob, & Varma. (2013). Financial Literacy among Working Young in Urban India. (1-27) W.P No.2013-10-02.
2. Atkinson, A. and F, Messy. (2012). *Measuring Financial Literacy: Results of the OECD/ International Network on Financial Education Pilot Study.* OECD WP on Finance, Insurance and Private Pensions, No. 15, OECD Publishing, http://dx.doi.org/10.1787/5K9csfs90fr4-en
3. Bahl, S. (2012). A study of advanced financial literacy among the working women in Punjab. *ZENITH International Journal of*

Business Economics & Management Research, **2**(12), 176-190. http://zenithresearch.org.in/1 retrieved on 30/03/15.
4. Beal, D.J. and Delpachitra, S.B. (2003). Financial Literacy among Australian University Students. *Economic Papers: A journal of applied economics and policy,* **22**(1), 65-78.
5. Bhushan, P. (2014). Relationship between Financial Literacy and Investment Behavior of Salaried Individuals. *Journal of Business Management & Social Sciences Research (JBM&SSR).* **3**(5), 82-87.
6. Bhushan, P. & Medury, Y. (2013). Financial Literacy and its Determinants. *International Journal of Engineering, Business and Enterprise Applications,* **4**(2), 155- 160.
7. Chatterjee, K. (2018). Financial Literacy Amongst Adult Male Slum Dwellers of Bidhannagar. *International Journal of Research in Commerce and Management,* **9**(3), 13-18.
8. Jain ,Rajeshwari.(2014) An Analysis of Income and Investment Pattern of Working Women in the city of Ahmedabad. *IRACST-International Journal of Research in Management & Technology (IJRMT),* **4**(6), 138-146.
9. Kothari, C. R. (1999). *Research Methodology: Method and Techniques* (2nd revised ed.).
10. New Delhi: New Edge International Pvt. Ltd.
11. Lusardi, A. (2008). Household Saving Behavior: The Role of Financial Literacy, Information, and Financial Education Programs. doi:10.3386/w13824
12. Moore, D. L. (2003). Survey of financial literacy in Washington State: Knowledge, behavior, attitudes, and experiences. *Social and economic science research centre, Washington State University,* Technical report, 1-62.
13. Tamimi & Bin. (2009). Financial literacy and investment decisions of UAE investors. *The Journal of Risk finance, 10*(5), 500-516.
14. Vijay laxmi and Maheswary, N.K. (2018). Identification of Factors Influencing Financial Literacy: A Theoretical Review. *International Journal of Research in Management, Economics and Commerce,* **8**(1), 89-94.

www.ingramcontent.com/pod-product-compliance
Lightning Source LLC
Chambersburg PA
CBHW050459190326
41458CB00005B/1349